P9-DCX-128

As the World Burns

AS THE WORLD BURNS

THE NEW GENERATION OF ACTIVISTS AND THE LANDMARK LEGAL FIGHT AGAINST CLIMATE CHANGE

LEE VAN DER VOO

TIMBER PRESS • PORTLAND, OREGON

Published in 2020 by Timber Press, Inc.
The Haseltine Building
133 S.W. Second Avenue, Suite 450
Portland, Oregon 97204-3527
timberpress.com

Printed in the United States of America

Text and jacket design by Matt Avery

ISBN 978-1-60469-998-2

Catalog records for this book are available from the Library of Congress and the British Library.

One of the benefits of fighting for a cause at a young age is that your ideas have no boundaries. It allows you to use your creativity and imagination to solve problems. We just followed our hearts and did what we thought was right. We didn't listen to anybody who was trying to tell us that we were too young or that our voices didn't matter.

XIUHTEZCATL MARTINEZ, *We Rise: The Earth Guardians Guide to Building a Movement That Restores the Planet* (with Justin Spizman)

Contents

CHAPTER ONE

October 29, 2018: A Crisis at Hand, a Nation Consumed

The sky tears open and drops out of itself. It is fall in Oregon, so the rain is not a surprise. The surprise is that the court is mostly vacant, the trial hasn't started, and nineteen of the twenty-one young people who are suing the federal government over climate change are standing on the steps having a protest instead.

They have come from all over the nation, from Hawaii and Alaska and New York, most traveling from landscapes already scarred by worsening weather. The youngest came from a barrier island off the east coast of Florida, where a red tide has just invaded the beaches and the fish are turning up dead. Another, from Colorado, has been hunkered down indoors with air purifiers during the worst fire season in the state's history. A third is living on a concrete slab in Louisiana, her family not yet able to replace their floors after a flood two years ago, courtesy of torrential rain. Collectively, they are litigating the deaths of fruit trees; the swallowing of islands; injuries to farming, fishing, and recreation; and the relentless worsening of wildfires in the West. They wear suit jackets and skirts and expressions of frustration and defiance.

Things do not work as they used to. That much is clear.

Philip Gregory, an attorney for the young plaintiffs, will soon clarify the rest. He approaches a podium midway up the steps to this courthouse in the Ninth Circuit and offers what the day's emcee terms "an update directly from the legal team."

He begins with something of a war cry.

"In August 2015, these young plaintiffs filed this lawsuit. This. Is. No. Ordinary. Lawsuit!"

His words are broadcast by speakers on both sides of the podium while the rain kindly slows its drumbeat on a pop-up tent overhead. The weather has been blustery enough that someone thought to strap the tent to cinder blocks, but the crowd seems mostly unfazed. Clad in rain hats and waterproof jackets, hoodies and damp sweaters, about fifteen hundred people are already shouting. Organizers have been pumping them full of funk music, and a tribal elder just called the plaintiffs "guardians of the earth." People are amped.

Gray hair swept back, Gregory describes the case as a constitutional one, brought by youth who are taking on the feds. "These young people this morning were supposed to walk into this courtroom and begin putting the climate science into evidence. But the federal government is waging a war defending the fossil fuel interests!" The crowd boos. One man behind me shouts: "Actually, I think it's the Koch brothers!"

"The federal government has taken their fight all the way up to the United States Supreme Court!" Gregory hollers. "And they're choosing the fossil fuel companies over these young plaintiffs, over the youth of America, over our posterity!"

This is his way of detailing why the Supreme Court just stalled the plaintiffs' case, *Juliana v. United States*, ten days before trial. It's a lawsuit that posits that young people—the plaintiffs, and future generations too—have a constitutional right to a stable climate. Its charge is that the government's actions to cause climate change violate their civil rights to life, liberty, and property. Not only that, but also that the government has known about the risks of climate change for decades and persisted in helping to cause it anyway, failing to implement its own plans to regulate greenhouse gases while subsidizing, authorizing, and permitting a fossil fuel energy system that worsens global warming every day.

The case is only one piece of a legal battle led by the Eugene-based Our Children's Trust, a public interest law firm that helps young people fight climate change. But it's a big piece. The only federal piece still standing. Whether now or later, it's the piece most likely to finish in the Supreme Court, making it one of the most important lawsuits of our time. What makes it especially unique is that the US government does not plan to dispute climate science at trial. Its experts have already conceded in documents that climate change is an urgent threat and that the United States is solely responsible for 25 percent of global carbon accumulation since the industrial age—more than any other entity on Earth. What looms for America is less another scientific debate, another fresh round of punditry, than an overdue public awakening. In a nation with dramatic influence on climate issues worldwide, this is *the* trial.

Now the plaintiffs stand beside Gregory against the glass facade of the federal courthouse, festooned as it is with banners: "Give Science Its Day in Court." "Let the Youth Be Heard." Theirs is not so much a formal flanking as a DIY ethos that has them scattered all over the steps. They are energized and at least a little bit angry.

Four are in imminent danger of being displaced by climate change. Six are Indigenous, battling for land and culture. Two are competitive skiers adapting to slush. And another duo were raised on farms threatened by the proposed path of a natural gas pipeline. To say they are being cheered by the liberal town of Eugene, Oregon, is to minimize the fanfare. Placards outnumber umbrellas. Vendors sell hot chocolate under tents. And police have blocked off the street so supporters, some from as far away as Alaska, can carry banners through golden leaves without fear of a traffic incident.

Represented as they are, they are plainly not the youth of Internet-troll lore, stage-managed by hand-wringing parents who coach them from beyond the frame. Never is that more obvious than in situations like this one, in which the plaintiffs command the crowd while their parents are left alone to make friends, always the first to look tired. While some of the parents are activists, too, and introduced their children to the cause, often it is the reverse. Either way, each young plaintiff joined the case on their own, heard about it from some corner of their own world. Whether the adults followed them more deeply into the fight is beside the point.

Gregory tells the crowd that what is happening to these young people is a lot like what happened with the youth in *Brown v. Board of Education,* when kids argued their strongest grievance against the government all the way to the Supreme Court, and the court stood up for them.

"We believe that this Supreme Court will have courage, will have the same faith in the youth of America in protecting these young plaintiffs' rights and will let this trial go forward," he yells, then pauses. The crowd is gamely filling these silences with whatever spirited thing tumbles out. This time they scream, "Yeah! Let's do it!"

The nation will be ill served, he says, if the plaintiffs cannot walk into that courtroom and testify how they've been harmed, and if their experts can't testify, too, to a judge who can craft a remedy for the environmental mess all young people now face.

"This is not just the trial of the century. This is the trial *for the future* of this century!" He finishes to wild cheers.

I've been a journalist for a couple of decades, much of the time focused on the environment, and this dust-up between these twenty-one litigants and the federal government is one of the most fascinating things I've ever seen. In its early days, *Juliana v. United States* was dismissed as a media ploy, a legal longshot, but it has defied expectations in coming this far. For the last six months, I've found myself captivated by these young people at the center of it and how far this case will go.

Some of the words Gregory just shouted are not his own. "No ordinary lawsuit" comes from a ruling in the US District Court for the District of Oregon on November 10, 2016, in which Judge Ann Aiken issued the first-ever opinion that people have a fundamental right to a climate system capable of sustaining human life. That this case could be what tips US action on climate change now seems like a serious possibility.

I am not the only one who sees this, and the carnival-like atmosphere only portends what's ahead. In a few short months, *Juliana v. United States* will belong to a larger youth call to action on climate, and these plaintiffs will move from mainly local renown to the pages of *Vanity Fair*, prime-time television, and celebrity selfies. That the lawsuit has its own podcast, and a documentary crew is among the media spectators in this crowd, are tiny details on the route of a fast-speeding train. On the courthouse lawn, posters feature each of the plaintiffs, and people are recording on cell phones, wearing *Juliana* patches held fast to their clothes with safety pins.

For the last three months, I have been meeting the plaintiffs and their families, background interviews to prep for trial coverage. I expected climate kids from affluent families, well-mannered crusaders with good teeth. None of them are what I expected.

Ten days ago, for example, Kiran Oommen (whose preferred pronoun is *they*) was in a coffee shop in Seattle when they got the news that their deposition was canceled and spontaneously cheered. Kiran was sitting in a window in the usual folk punk ensemble, dark jeans and high hair, metal studs in a denim jacket with patches and cut-off sleeves. They were not hoping for the trial to be canceled. Not at all. It was just that the thing Kiran hoped for more than anything else right then was a weekend. Two days in which they weren't working or studying or organizing a rally or a small concert in the kitchen of the rental house that doubled as a music venue.

Pharrell Williams's debut solo was incongruously playing on the sound system when the text buzzed in on Kiran's phone. "Deposition's canceled. Ha!" Kiran clapped. Their mood, which hadn't been bad, brightened noticeably nonetheless. "That's wonderful," Kiran said, and exhaled like someone who had just been told they could keep their kidney after all. "Well, that's good. I can work on Monday, and maybe I'll actually pay my rent at the end of the month."

It was a Friday, and the night's bands were already arriving at their house. The rest of the collective and some of Kiran's five roommates were busy setting up, but suddenly there was time to spare where lately there had been none. Days were waking at 6 a.m., changing lightbulbs and doing groundskeeping for the University of Washington, sucking on what seemed to be an unhealthy amount of smoke at the peak of the fire season, before tackling all the rest—school and organizing and music too. There were lots of days when Kiran headed home fourteen hours after leaving it, falling-down tired. And though Kiran wasn't excited about being interrogated by the federal government, they were less concerned about this interrogation than about the mayhem of just fitting it all in. And because Kiran was arty, needed time to make things, they had been feeling off-kilter, as if a reset button could not be pressed. But a weekend. It would fix everything.

Flash forward, and now Kiran stands on the steps of the courthouse, wearing the kind of finery that is appropriate for a courthouse. Gone are the usual metal studs and pins, the sewn-on patches. "Diverse" is the light word for how to characterize the personalities in this group. Kiran knows they, meaning Kiran specifically, are the reason for the indirect but polite email about how important it is for the plaintiffs to not get arrested. They were an organizer from early days—their father a teacher, their mother a minister for the United Church of Christ that still eagerly claims them, nonpracticing though Kiran is—but on some days, their activism is more radical than is easy for everyone else to stomach.

Kiran doesn't believe in top-down structures. In placing people on pedestals. In looking at other activists as anything other than fellow crusaders. They think that when you elevate people to such posts, the result is that those people tend to get away with things. Kiran doesn't want to get away with things. Kiran wants to work alongside everybody else for this cause they care deeply about.

So when Kiran was asked if they were interested in joining the lawsuit, they were already veering from the mainstream, an activist about to graduate from high school but already involved in direct action. Kiran had never heard of people suing the federal government. The very idea of it "just sounded so absurd to me that part of me was just like, 'Oh, this is hilarious. I would love to say that I'm suing the government.'" So Kiran signed on. Then there they were, on the steps of the federal courthouse, suing the government. Many of the people Kiran knew would not go in for this: changing the system through the system. But Kiran believed that if you truly mean to make change, to agitate, you rattle all the doors. Even the front door.

Now the front door of the courthouse is directly behind them. It seems to be, at least metaphorically, locked for now.

In another few minutes, Julia Olson, the plaintiffs' lead attorney, a passionate speaker and much less a yeller than Gregory, takes to the podium to deliver a clear-eyed speech aimed less at the crowd than at the Supreme Court itself. It is a speech that calls out a weekend of horrific violence in Pittsburgh—yet another civilian massacre, this time at a synagogue—as a way of highlighting America's track record with ugliness. Slavery, colonization, land theft, and internment. She notes how the Supreme Court has been an arbiter of much of it. An effective check on rogue power, a force of light in dark times.

"One of our most enduring traditions as a nation has been one that's been emulated around the world. And it's our system of checks and balances. It's a judiciary that can hear the strongest grievances against our government, against our political branches," she says from the podium, shoulder-length curls damp and curling tighter. "Even grievances that question and challenge long-standing practices of our government that betray our children."

She quotes justices John Roberts and Sandra Day O'Connor, in moments when they defended their own independence and the import of children as citizens of democracy.

"Now the court has, from time to time, erred and erred greatly," Olson says, like the time it defended the government's decision to jail Japanese Americans during the Second World War. When the court has made these shameful, grievous errors, she says, "it has been because the court yielded to political pressure."

She doesn't say so outright, but this, in a nutshell, is what people fear will happen. The *Juliana* plaintiffs, ranging in age from eleven to twenty-two, have just collided headlong with the Trump administration's steadfast denial of climate change and its deep ties to the enclave of wealthy one-percenters who got that way through fossil fuels. It is not an ideal time for the court to tend to politics.

These moments in this case, when humanity seems to be sliding ever more quickly toward an abyss—carbon dioxide in the atmosphere zipping past 410 parts per million and nothing to stop it—I try to imagine how this story will be resurrected and retold. Thousands of years from now, will an alien species find a lone library still standing on high ground? A digital tomb stuffed with urgent hashtags in Yucca Mountain with the nuclear waste? Maybe it will be a cache of YouTube videos, catapulted into space or to Mars on one of Elon Musk's rich-guy rockets. What thoughts will they have when they see this mess, poke at our fossilized bones?

Here are mine: As I follow these plaintiffs for the next year, these days to test whether the American courts can protect the young from an imminent climate breakdown, they could have kept the world from burning. They didn't. And this story of the youth at the heart of this case, the *Juliana* twenty-one, trial or none, it's a story people ought to know. Not because there are times when kids can stand to learn a lot from an adult perspective. This is definitely not one of those times. But because if everyone in this country could see what I see, watching twenty-one young people question the value of their own lives against the entrenched politics of a government that threatens them, maybe this nation could do a better job of answering the question they have brought to the courts, to us all: Do children have a right to inherit a livable planet, and if so, is the government obliged to protect it?

To see the world through their eyes, see this question take shape, it is a heart-wrenching thing.

Since this lawsuit is her namesake, Kelsey Juliana does not have the luxury to look away.

Two months ago, she leaned her bike against a picnic table and sat down on the bench at Amazon Park in Eugene, a hundred-acre park just south of downtown, a place with miles of trails, a swimming pool, and lots of kids. Her parents' home isn't far, a modest bungalow with a less modest garden, so she used to play lava monster here back when the playground was mud instead of turf.

Nobody knows the name Greta Thunberg yet. Right now Greta Thunberg is a frustrated fifteen-year-old beginning solitary protests in front of the Swedish parliament, garnering little media attention and not much support.

But like Greta, Kelsey has been a climate activist from an early age, since she was ten. That was twelve years ago. Lately, she notices that something is different. The buzz around her trial is intensifying. A lot.

There are about a dozen kids on the swings behind her, diving for the green turf carpeting the ground. Their playful sound is like an echo from a carefree past. It is very distant for Kelsey. "I try to not dwell on the fact that two hundred species go extinct a day, and even if we were to stop fossil fuels right now, we'd still be feeling climate effects for up to one hundred years," she says. "A lot of other places in the world . . . the issue of climate change is an issue that exists. It's not like, 'I don't believe it.' Or, 'Do you believe it or do you not?'. . . The problem with the US is that it's a political issue. Period. And it's not about our livelihoods. It has nothing to do with what's accessible, what's right or wrong, what makes sense. It has to do with: 'What party do you stand with?' . . . That's why this case matters. Because it takes away the politics. It's not about Trump. It's about a system within our government that meets the needs of industries instead of people."

If Kelsey has a superpower, I will learn it is this: she can take the temperature of things. She can read not just a room but the collective noise of what is rising in the form of environmental zeitgeist. Later, I will notice the way she can conceive a thought—almost casually, in the middle of a conversation, a meal—and seize on something brewing in the culture, articulate what otherwise feels intangible. It seems she holds these reflections all the time, her mind a mirror of how the world cradles this cause. When she talks, gives speeches, and goes to classrooms to work with kids, she pushes these ideas back toward crowds and builds on them at once, underscoring values, advancing them further. She does this with no preparation, no notes, no rehearsal. It is more a way of being. And when she speaks, it is like she's been listening hard and what comes out of her mouth is a translation of what she's heard.

This day, her long brown hair with its touch of auburn seems in place no matter how it lands, and the sun shows on the tiny hoop in her nostril. She has large blue eyes that somehow never eclipse her face, fair skinned and oval. The air is warm, the sky a clear blue, and the leaves in the trees have a late-summer hue, a promise of fall.

Most of this work is drudgery, the highs fleeting, she says. The bits she loves are often few. In between, she practically needs a secretary, her days so full with calendaring and travel planning and appointments, plus the banal tasks of just being a person. "I'm still trying to just do my chores and, you know, wait on the phone with Equifax for forty-five fucking minutes," she says, trying to check her credit.

In other words, she is not naïve about what she can expect in return for the investment of her time, which is frequently nothing. Sometimes worse than nothing, like when she is watching interviews of herself looking drained and unhappy and not being able to imagine a Friday night with friends instead of answering email and trying to just keep going—advocacy has just that much pull on her. But the trial? With her keen ear for how much noise this work could make, what she hears this summer is the volume pitching up, a change in how the space around her is beginning to feel. It's what she's been waiting for.

People are already talking about showing up to court on opening day, and what she is telling them is: "Yes, show up. But show up that day and one other day too. Blog about it. Post about it. Bring a friend." "I don't want it to be a resting place," she says. In other words, not like a march. A thing you put on your calendar, get excited for, turn up to, and then forget. The trial will take weeks, months. She wants people to see it the way she sees it: as a day to start whatever comes next, even if all there is to start is the marathon of keeping up the pressure. It is weeks of spotlight, months even. And what people do with that is important. Maybe it is all that's important.

Kelsey is ready. "I hope that it fires up and ignites something," she says. She wants it to be more mutiny than march, this thing rising up in her name. A shift to define all the ones that will follow. She has started to dream about the date. With the kids still squealing on the swings behind her and her bike against the table, ready to carry her to a meeting with a friend, she describes how October 29 came to her in her sleep, the big OCT—also the acronym of her lawyers' organization—dancing toward her and a feeling of excitement building until she woke and it spilled into her life.

Kelsey is like every other plaintiff in the case this way: in the months and weeks before trial, she is preparing to step into the spotlight and hold it in a way that can spur real change. All the ingredients are there. The Parkland shooting has just happened, an active shooter killing fourteen students and three teachers at Marjory Stoneman Douglas High School in the town in Florida, and the survivors are galvanizing similar social change in the aftermath. Emma González, who "called BS" on political reaction to gun violence after the massacre, has launched the Never Again Campaign. And David Hogg, her friend and activist cohort, crushed Fox News host Laura Ingraham with a swift boycott of advertisers after she mocked him about a college rejection letter. Together with other of the Parkland youth, the duo are stumping for gun safety laws across sixty cities in sixty days, and creating the political climate that will soon lead to the passage of seventy-six state laws and increase youth voter turnout by 10 percent in November.

As this happens, the thing that is coming into view for Kelsey and many of the other plaintiffs is that they stand at a very similar nexus: a place where political inaction and youth frustration are building high enough to boil. While young activists are taking to social media and building voice and momentum around important things that lack both—gun violence, Black Lives Matter, Standing Rock, climate change—*Juliana* is beginning to trend.

It helps that Our Children's Trust is putting a lot of effort into ginning up the social media mill, dedicating staff to riding the #YouthvGov hashtag with trial updates and picture quotes from the plaintiffs. Bankrolled by left and the moneyed left, including celebrities like Leonardo DiCaprio, the *Juliana* case has in its arsenal several not-so-distant connections to celebrity and the attending audience. By proxy, the amplification that could come with Hollywood once Hollywood gets hold of the case is not far off. And in this, it seems *Juliana* is a sleeping giant slowly waking.

Big names like Ellen DeGeneres are already pushing the news through their channels, signaling an upcoming media pile-on. Nearly every teen glossy and trendy digital news magazine has had at least mention of the case, if not a profile of a plaintiff or a write-up of the litigation. And once these outlets are on the bandwagon, they tend not to stray. Briefs and updates. Tidbits of legalese in anticipation of the next. Which means that if SEO and social media are about to be as weaponized as the amicus brief, the case could be propelled onstage and onscreen in ways that even Johnnie Cochran, master of the courtroom media spectacle, could never have fathomed.

Daily news outlets and wire services are already betting on it big, with Reuters (that's me), the *New York Times*, the *Washington Post*, and others planning trial coverage that includes extended stays in Eugene and a healthy dose of pretrial jockeying for space in the courtroom. The documentary crew has also been following the plaintiffs for years, recording what is likely to be history one day. And for the Our Children's Trust podcast, several of the plaintiffs are carrying USB recorders, memorializing their daily thoughts to a team of audio producers.

Now, the news cycle is poised to dive into the case. And in these ways, the *Juliana* twenty-one have the perfect luck of both timing and age. Most are too young to vote. And they filed their case against the

Obama administration, not Trump. Both facts mostly quiet the charge that they are politically motivated. That they are too heartfelt, too real, too young to smack in public is just a bonus. In a few short weeks, the media will turn its bright eye toward them and the stage will be theirs.

Kelsey wants them to really own it.

Some of them already do.

Xiuhtezcatl (shoe-TEZ-caht) Martinez, namely, has had a microphone in his face since he was six. The footage is on YouTube—videos of him standing on an open-air stage rallying tie-dyed crowds over the fate of the planet as a pudgy, squeaking little kid. Later, he would become the youth director of Earth Guardians, spurring hundreds of chapters in fifty countries, and speak to the General Assembly of the United Nations. He has already been a two-time guest on Bill Maher, has appeared on *The Daily Show*, and lately has been shirtless in a Vice i-D brand profile and sporting new kicks in an Adidas campaign. If you know him, know what he looks like, it is not at all odd to see him while you are moving around in the world, his long dark hair on some billboard or in yet another photo in yet another magazine. His hip-hop career is accelerating from where his leadership post with Earth Guardians left off, and he is certainly a staple in a media diet if not yet an icon.

Which is what makes a thing that happened the summer before the trial especially odd, and perhaps lends credence to Kelsey's notion that the trial could ignite something bigger than the climate cause had yet. It was a simple, seemingly benign exchange on Twitter, the opposite of viral.

August 12, a tweet from Donald Trump: "Xiuhtezcatl's new EP is fire, so hot it's making me believe in climate change."

Xiuhtezcatl's response: "#NotFakeNews" and an emoji pointing to a link to the EP.

Actually this wasn't a real exchange, it was a meme. But I couldn't tell, a fact that underscored the widening gap between expectations and actual events, especially where the president was concerned. That this tweet could be a meme or just another day on social media, Xiuhtezcatl caught up in the momentary glare of a president who can't resist a good digital row, was indiscernible. Despite the fact that Xiuhtezcatl was seventeen years old at the time and the president was a bona fide senior citizen, the specter of the impending trial made this digital tête-à-tête especially possible. Donald Trump and Xiuhtezcatl were not merely contemporaries in pop culture stardom. They were adversaries, too, with Xiuhtezcatl prominently named in the lawsuit alongside Kelsey and angling to face the Trump administration in federal court.

It was not out-of-bounds to think Trump had bestowed on Xiuhtezcatl the dubious distinction of being held within a close but albeit large sphere of presidential frenemies. Spats with Rosie O'Donnell were the barometer of the day, proof that celebrity was all it took to ascend to this digital club of executive sparring partners. With Aztec looks that played as well in a tuxedo as in jeans, Xiuhtezcatl was fast becoming a favorite to name-drop among name-dropping environmentalists who had perfected the pronunciation of his name. His message still about saving the planet, it seemed he might be worthy of watching, of tweeting about and perhaps skewering later by a president with a gift for manipulating public perception. It seemed reasonable that Trump could see these young plaintiffs rising into the epicenter of a weak spot. The president was already failing to control the message around the Parkland shooting and to lure the spotlight away from the young people who were ambushing his status-quo agenda on guns, poised as they were to set law and precedent in state after state. Now this hip-hop artist, this spokesman for the climate crisis, was coming into view. Maybe he was an opponent worth besting first. Maybe all of them were.

Every day the rules were being remade by a leadership that did not like the old ones. The truth was suddenly malleable, the press an enemy of the people. Russia was in, Europe out. Coal was cool again. And the authority of the courts was being eroded by insults and appointments, with Trump alternately nominating federal judges and slamming the court system overall, calling it "slow," "political," "a joke," and "a laughingstock," even blaming it for terrorism and charging that judges were some of "the most dishonest people in the world." He was particularly frustrated with the "outrageous" Ninth Circuit, in which *Juliana* was unfolding, complaining that his opponents chose that "broken and unfair" venue to procure the decisions that most vexed him. He openly mused about how to break up the court and said he was considering proposals to do so. All the while, it was as though the playing field were being shifted, the lines redrawn. It was hard to trust in expectations.

Who knew what spectacle loomed. These possibilities underscored the peculiarity of the hour: a crisis at hand and a nation consumed with whatever strange thing its president would next do.

The next time I see Kelsey we are at the rally on October 29, and it's as if the whole planet has shifted from where we last stood on it, the crowd roving behind her, moving from the speeches on the steps of the courthouse to a stage across the street. After the attorneys have spoken and the plaintiffs are taking turns at the podium, she is on her crowd high, zipping between the people who want to talk to her and the other people who want to talk to her. She stops to give me a comment for the story I will file imminently and says, "It just occurred to me I don't even know where my phone is. Or my purse. People are trying to track me down and I can't even track my own shit." She doesn't seem worried. Though we are still in the street in front of the courthouse, it feels like a party in someone's very large yard.

Juliana had been set for trial in the US District Court of Oregon since April, but the Department of Justice had asked the Supreme Court to throw the case out of the lower court as the October 29 start date neared, claiming the trial was too burdensome and the court lacked the jurisdiction to call for such a sweeping review of executive branch powers. Attorneys also had told the Supreme Court there is no such thing as a constitutional right to a stable climate.

At first there wasn't much to fear—this had happened before. The government had already asked the Supreme Court to toss the lawsuit back in July, and had three times asked the Ninth Circuit Court of Appeals to do the same. The legal maneuver—called a writ of mandamus, an emergency writ—is a heavy-handed one, a rarely used weapon wielded by persecuted parties to petition a ranking court for relief from a subordinate court, one that's harassing them. But despite its more tempered uses elsewhere, this was the fifth time the Department of Justice under Trump had tried to use an emergency writ to argue its way out of the *Juliana* trial. The behavior underscored the president's commitment to reducing the judicial branch's powers to check his administration. The plaintiffs had won every time.

Legal experts would later call the delay a Trumpian attack on civil procedure and the courts, nothing less than a siege. In the sixteen years before Trump became president, the writ had been used just eight times—in emergencies during both the Bush and the Obama administrations. Since Trump's inauguration, it had already been used twenty times, half of those the month of the would-be *Juliana* trial. The tactics were replicated in other cases close to the president's agenda: the Muslim air travel ban, the ban on military service by transgendered people, the challenge to the rights of young Dreamers to remain in America, and the call for a citizenship question on the US Census.

This kind of unprecedented aggression—using a stacked Supreme Court to subtly pressure lower courts to dispense with litigation—had the destructive side effect of fatiguing attorneys and plaintiffs, scrambling court calendars, and delaying trials just for the sake of delay. It was hard to know how it would end.

The last stay had dragged on for a year, and the jitters are palpable. Everyone seems to know that the Supreme Court will rule in the plaintiffs' favor as it did unanimously in July. Still, the controversial swearing in of Justice Brett Kavanaugh three weeks earlier had the press chasing legal experts, some of whom now presumed the Supreme Court's rulings would be different. With the media awash in reactionary what-ifs, few can escape the feeling that although all signs point to a trial that is imminent, the signs are not to be trusted.

I ask Kelsey how she feels about the delay.

"The delay is just, it's like fuel to the fire. It makes the government look desperate in their self-interest, which is exactly what it is, its desperation to preserve their self-interest of continuing a system of the usual, which is a system of exploitation and corruption," she says. "It makes us just mad as hell and ready to get in there even more." Yes, she wants to be in trial today. So do all the other plaintiffs. But by the time that happens? Now she figures they can double the crowd, work out the kinks. Maybe they *needed* a test run.

Some version of this plays out throughout the day: plaintiffs angry about the delay who see their government as an aggressor.

"It's definitely indicative of how the government has been acting this entire time," Vic Barrett says, the plaintiff from White Plains, New York, whose grandparents' Honduran home is being washed away by sea. "They have no intention of looking at the evidence that's right in front of them, so it's really disappointing but not surprising that they are going to such extremes to prevent the trial from happening."

The rest of the time the plaintiffs are just glad to see each other. There are hugs and huddles and—during breaks in the speaking and the performances—a lot of shoulder-to-shoulder pacing. They have forged unique bonds around climate activism and the bullying, stress, and blowback some of them have endured since joining the case. Being a plaintiff is, for many, more intense than they'd expected.

The initial targets of the lawsuit were President Obama and Obama-era appointees, the initial aim to provide a legal framework to an administration that showed some interest in reining in climate impacts. And at a time when entanglements between energy interests and the federal government seemed to be shifting, however entrenched they still were.

But then Trump happened. And instead of facing President Obama and appointees like Steven Chu, the former energy secretary who'd once conceived an alternate economy based on plant glucose, the plaintiffs are up against a president who stumped for coal in West Virginia and a cabinet of climate change deniers and former fossil fuel industry executives. Fiction could not have crafted a better cast of villains. There is Secretary of State Mike Pompeo, who has publicly waffled on the existence of climate change. Then Secretary of Energy Rick Perry, who is spearheading efforts to increase coal production. Secretary of the Interior Ryan Zinke, who has increasingly opened public lands to oil, gas, and mineral extraction. And then there is President Trump himself, who has set benchmarks for drilling and fracking beyond what markets can even bear; stripped the EPA of its regulatory powers over air, water, and cars; gutted budgets and international agreements and muzzled scientists; even scrubbed the words *climate change* from public documents, websites, and policies. Other cabinet members in Trump's fickle administration have similar leanings—former EPA head Scott Pruitt once told an interviewer that "humans have most flourished during times of warming trends"—and have since come and gone as *Juliana* defendants amid fleeting tenures as government appointees.

All of them are worth shouting about on the steps of a courthouse.

While the shouting starts to give way to music, I do as journalists do and write the day's events down, synthesize them, and feed it all to the machine.

I have not yet digested the news that after today I will be suddenly unemployed. I'm still in a frame of mind I adopted some months ago when I decided to drop a longtime contract gig to focus on covering this case. At the time, the trial was still ahead and I'd been salivating over the court record for months, meeting the plaintiffs in the regions where they live. *Juliana v. United States* was the Scopes trial—the 1925 trial that let evolution be taught in schools—of my lifetime. It was a trial to command the news cycle and captivate millions. I decided I was going, then started shilling for commissions—emailing editors and drumming up assignments. Suddenly editors who were never interested in my work were very interested in my work. It was the plaintiffs, of course, who drew them. And this climate crisis that we could dimly see was boiling to frustration among the young.

Through the speeches and the note taking, the sidebar comments and the music and the cheering, I am only barely beginning to realize that the trial will not go ahead as planned, and that I am unemployed, zilch for commissions, living alone in the woods with a hefty cat named Larry and some very mean raccoons. Instead of panicking, I find myself distracted by the thought that has been on my mind for days: that tens of thousands of plant and animal species will be facing extinction within the year thanks to human activities and global warming. And even though the raccoons are faring well—proliferating even—as the earth's temperatures warm, American raccoons are moving north into Canada and raccoons in other parts of the world are leaving hot zones in favor of cooler ones. They are good at adapting—eating from the cat bowl and so forth—but that is not the issue. The issue is that the raccoons are being labeled "invasive," which is what the humans call you right before they start to kill you. Or send you off to Trash Island like in a Wes Anderson movie. And now thanks to plastic, Trash Island is a real place, lately the size of Texas. It even has a name—the Great Pacific Garbage Patch—and with this moniker has seemingly joined a league of continents. I imagine if we could ask them, the raccoons would be sick of this lens—the one through which all the rest of nature bends to human invention and will. I am sick of it too. It's why I wanted this trial. Why a lot of people wanted it.

Some of the plaintiffs have gotten hoarse making speeches, and the youngest, Levi Draheim, being the most energetic and the least equipped for the weather in a shiny blue sport coat, expends his angst by pedaling the bicycle-powered sound system under the pop-up tent. In short order, Kiran takes to the stage and plays a mandolin to guttural, self-styled lyrics about the case while Levi and Kelsey lose themselves in the dancing. Blue sport coat next to green blazer, busy black pants and shoes.

Three years later, now we had a date in court.
That doesn't mean we've won,
no this fight is never done.
So I hope that I will see you in the streets.

In between are poems, a two-part spoken word performance, an oral presentation by a class of fourth and fifth graders, and words from the mayor about how several of the local plaintiffs cut their activist teeth by lobbying her for what's now the nation's strongest local climate ordinance. At points, the Indigenous peoples in the crowd bring their own mojo, calling "Native Americans are here—Ah-ho!" so that the response call "Ah-ho!" drifts on the day like the contact calls of birds, support for the plaintiffs with Indigenous roots.

Everybody talks about how important it is to show up, to keep showing up. So all day long, they show up.

Jacob Lebel, one of two plaintiffs who are rural farm boys from southern Oregon, steps to the microphone in a suit jacket and ponytail and says what a lot of the plaintiffs seem to be thinking. That he isn't deterred by Trump. That this isn't what he'd expected, but it is where he needs to be. "The wildfires around my farm in Oregon keep getting worse, the winters keep getting warmer, the salmon keep dying, the glaciers in Nathan's home keep melting, Jayden's home keeps flooding, the seas keep rising, and our politicians keep *lying*."

His voice rises on that last word, and it seems like everyone cheers. They are activists from climate action groups, local hippies, and children who marched from nearby schools. Six of the plaintiffs are from Eugene, and the community has come to think of the case as its own. It's an important thing for a community to hold, so everyone tries to hold it and overlook the fact that instead of going to trial, the plaintiffs are lining up for speeches and photo ops.

It is a bold show for a group of young people who've been bullied out of their right to a trial. And it reverberates around the world. Thanks to connections with 350.org, supporters rally in more than seventy venues—holding demonstrations in every state and in Stockholm, London, Amsterdam, and Uganda within the week. While young voices take to microphones and banners wave, social media and the rest of the media light up with the noise. All of it buoys the plaintiffs, who wear bright faces and cheer themselves on instead of lamenting that they are having a protest instead of a trial. They make firm statements about how they will get their day in court, how this isn't over, how moral imperative and the state of the planet demand that they be heard.

It's hard to tell, though, how many of them really believe the court will someday hear them.

In the days afterward, after the ice cream wears off, and the movies and the hangouts are no longer sufficiently distracting, there are tears. In between, an Airbnb fills up with snack food like an episode of MTV's *The Real World* while the out-of-town plaintiffs wonder whether to stick around or go home. No one knows how long the delay will be. Or if it will turn out to be permanent.

What begins today is a tedious wait that will drag on for years. Through the next one, the urgency of the plaintiffs' question will only become clearer.

CHAPTER TWO

"The Country That We Want to Live In"

What the plaintiffs want—Kelsey and her twenty co-petitioners—isn't impossible to do. In fact, it has already been done by hundreds of Dutch people in the Netherlands. Except in this country, it's harder.

Jacob explains as much to me in August, two months before the October 29 trial date, in the kitchen of his Oregon farmhouse. He's dressed in white pants and a sweater, an outfit that seems for my benefit. Other days, he looks so much the part of farm boy in cargo pants and flannel that the documentarians following *Juliana* can't ignore the ethos he tends to provoke. These images—clips of Jacob petting a cow, releasing geese from a pen, hefting a bale of hay while whistling to a horse—have become signature to the case itself. When we meet inside his home, a place of smooth woodwork and books, his soft, Quaker ways are more pronounced. A chalkboard on a piano holds a collection of words akin to prayer: Help. Fight. Discipline.

"We're suing all the agencies," Jacob says, hair pulled back in the usual ponytail. Then he lists them. Ten agencies and offices of the US government. "Department of Interior, EPA . . . "

What the plaintiffs seek is a remedy. They want a judge to order the agencies to "swiftly phase down" carbon emissions, and develop and implement a national plan to stabilize the climate. In the Netherlands, nine hundred plaintiffs succeeded in suing the Dutch government on similar grounds in a case called *Urgenda Foundation v. The State of the Netherlands*. In that case, a district court ordered the nation to reduce greenhouse gas emissions by 2020 to at least 25 percent lower than 1990 levels. Carbon accounting, redress planning, and implementation are already under way, even while appeals—appeals that will ultimately fail to overturn the ruling—are pending.

What's worked in favor of the *Juliana* plaintiffs so far is that the district-level judges haven't been shirking the authority to rule similarly. "The judges are very clear. And they don't take any bullshit," Jacob says. When the government argued that the courts could not tell them what to do, Judge Ann Aiken in the District Court of Oregon reminded everyone that the courts have a long and healthy history of telling governments exactly what to do, throwing down the authority to mitigate on behalf of the oppressed. Prison conditions. Desegregation. It was exactly the kind of Ninth Circuit back talk that Trump loved to complain about, arguing that liberal judges are taking over the courts, "a disgrace," and so forth. Still, Trump had just been dropped as a named defendant in *Juliana v. United States* to avoid the drama of his attention to it, even as the plaintiffs continued to sue the office of the president.

Jacob sits at a hefty wooden table lined with benches, the nearby walls wrapped in mason jars filled with grains and beans and nuts, and considers the merits of watermelon. More specifically, this watermelon, which he assesses as substandard, then apologizes for not having grown, clarifying that the farm's watermelons are coming up but not ready.

"It's kind of not as perfect as I would like," he says, meaning the watermelon.

Then he explains that the plaintiffs' goal is a "climate remediation plan" for America, an oft-repeated phrase that, till now, breezed over my head like swifts en route to a sunset chimney. I don't know what he means, what it would look like for the federal government to actually fix its part of the climate crisis. But then he breaks it down. Piece by piece. How each branch of the government would assess its climate impacts and set goals to roll them back, then reduce carbon, if the court ordered as much.

"It's not just about stopping the emissions. It's about policies to draw down carbon," Jacob says. "All those agencies have the power to do that—to set policies, set regulations, set public examples."

In a postvictory world, the Department of Agriculture would assess methane release from, say, billions of shitting cows, then redefine best practice in ways that set goals for emissions. The US Forest Service would calculate how many trees it takes to offset a scientifically necessary amount of carbon emissions and lease timber accordingly, maybe even plant more. The remedies are out there. And the plaintiffs want feasibility studies of the economic and technical requirements to decarbonize all of our national systems; in other words, they want a plan.

From each agency targeted in their lawsuit, they want an accounting of all the carbon that agency is responsible for. Like emissions from the five hundred military bases and three hundred thousand vehicles under the purview of the Department of Defense. And leases for coal mines and oil and natural gas wells on public lands. And the promotion of fossil fuel imports and exports, plus permitting for new energy facilities and pipelines. The impact of leases for offshore drilling. Of carbon standards for power plants. Of air travel. And they want the government to determine how all of that can be balanced against the carbon contained by energy efficiency programs, by ocean and atmospheric protections, the trees on federal land, fuel efficiency standards, and whatever international influence the State Department could have on those same things elsewhere in the world.

"One of the nice things about it is if we succeed, all of the agencies will be forced to work together, which they don't do," Jacob says. To have the departments just take a full carbon accounting of the activities of the United States? "Even if we get just that, it would not do much but it would be amazing."

Implementing such a plan would touch everything we can imagine about the way we live. How our food is grown. How we travel. Our energy systems. Fuel prices. The way we finance things and how we understand money. Even the military's planes and jeeps and tankers. And yet, it can be done. Redress planning and implementation in the Netherlands is proof of that. And even while there is deep debate about whether the Dutch plaintiffs compromised too much, settled for what was possible instead of what is necessary, the Netherlands is moving toward being a carbon-neutral nation, even as appeals are pending. It can happen here the way it happened there, and with careful planning and a lot of math, nothing has to break along the way.

Jacob explains this simply enough that it leaves me wondering: *Why is this so fucking hard?*

But it is hard. In a legal sense, it's hard because the Dutch courts are a different beast. International human rights have long fed into their interpretation of the constitution. Plus, the government's obligation toward the health and welfare of its people is explicitly written, not inferred through rights to life, liberty, and property like in America.

It's also hard because America is dug into this system of fossil fuel energy and car culture and monocrop farming and billions of shitting cows, as is our economy. The strongest nations in the world depend on an enormous and cheap supply of energy. And the folks who benefit from doing nothing about the environmental consequences of that dependence spend an extraordinary amount of money and time convincing Americans that to do nothing is the correct response. As part of that effort, they help put people in positions of power whose very intent is to do nothing, and those people in turn make do-nothing policy and criminalize, at times, the people who fight the status quo. Now, time is running out.

Global temperatures have risen nearly half of the 2 degrees Celsius that scientists worldwide agree will melt the earth's ice, expand oceans, and boost the level of the sea by half a foot or more by midcentury. All this will intensify storms, reduce snowfall, and rearrange the hydrology of the earth, causing drought and fire, heat waves and disease, food insecurity and famine, and paving the way for mass migration and human conflict like the world has never known.

To meet Jacob on his farm is to see—really see—how all of it could be different. When I meet him in the summer of 2018, he's just made a trip to Arcosanti in Arizona, the experimental city that's redesigning the way people live on the earth. Begun by Italian architect Paolo Soleri in 1970, it is the kind of marriage of architecture and ecology that can send Jacob's head on a twirl. It's a place of taller buildings and no cars. East-to-west structures made from cement cast in earth and open, crescent-shaped spaces that minimize wear and tear on the landscape. The blueprints—spread on a massive conference table by people wearing the right kind of gloves—with their vaults and apses look like the blueprints of the Starship *Enterprise*. They are epic, exhaustive, so unlike anything else that they lie just outside the edge of the imagination for many.

But for Jacob, seeing Arcosanti and attending workshops there was a way of imagining the future of his family's farm. For more than twenty years, his family has been proving out the concepts of biodynamics; the idea that people can live off the land, and quite comfortably, without making a wreck of it. After years of searching for the right patch of earth, they found this piece of land and emigrated from Quebec to Oregon for it. Now, it is his dream to test what they've learned in harsher and harsher conditions.

"One of the more basic principles of biodynamic farming is to treat the farm as an organism. As a self-sufficient, self-contained organism," he says. As in manure from the animals feeds the soil, feeds the gardens, feeds the people who feed the animals.

He says this on the winding road to the vegetable garden, dusty in the summer heat, as we pass peaches and nectarines, ascend the hill into apples and pears. There are petite 'Babydoll' Southdown sheep, bees and goats and a couple of ponds. Elsewhere are Tamworth hogs, Jersey cows for milk, and Scottish Highland cattle for meat. There are rescue horses of mixed breeds and chickens and ducks and geese too. Sometimes there are turkeys, quail, and guineas. All for the eggs and the poultry.

This farm-as-organism is in a constant state of evolution. Two decades ago, a groundbreaking garden on the original 170 acres was cut from hand tools out of 6 feet of blackberry bush. First the garden, then a few animal corrals. The farm is 600 acres now. Jacob says it took twenty years just to get the gardens, the animals, and the orchards up to speed. In between, his family added land and machines until Jacob went from tending the animals as a child (he could honk at the ducks so that they lined up in a row, and sheep came running at his bay) to managing the garden and a small produce business before taking over the business of the machines. In the last month, he has fixed a forklift, a fire truck, two excavators, a sawmill motor, and a weed whacker. So his life is not so much absorbed with the climate cause as it is stocked with hydraulic fluid and engine oil and antifreeze. Some days he spends much of his time just trying to sort out which part goes with what machine.

Learning has just been like this for him. He figures it out as he goes along, often quite well. So it was that school was a thing that also happened on the farm, squeezed between all the chores. He was homeschooled until the end of eighth grade, passed the entrance exams for college, and earned an associate's degree by the time he was nineteen, pecking away at it along with all the rest. As he went from a young boy who played pied piper to farm animals to a young adult who is suing the federal government, the farm grew from his family's passion to a fully realized thought experiment.

This place is what gives Jacob standing in the *Juliana* case. All of it is threatened by the changing climate and the increasing encroachment of the natural gas industry.

For the past three years, Jacob and the other *Juliana* plaintiffs have been proving this standing in court. Proving they've been harmed. Proving that this harm is unique to them and not a thing that everybody has suffered. Proving that the US government has had something to do with it. In other words, tracing the harms in their lives to climate change, to carbon emissions, and proving that it is possible for the government to help fix things. The government, in response, has pointed to things outside its control. Other nations. Third parties. Businesses and, ostensibly, individuals with lawnmowers and cars and air conditioners. Through the Department of Justice, the Obama and Trump administrations have both argued they aren't solely responsible for the changing climate and that our nation wouldn't be able to address the problem alone anyway, pointing to China's emissions and other nations' too. They have pushed back against the idea that the United States could ever affect any of it. Plus, this question of the courts telling the executive branch what to do? That is improper, un-American even, as far as the Trump administration is concerned.

"This is the first time in human history that we face a challenge like this on a global level," Jacob says. "As we get overpopulation, as we get more greenhouse gas emissions, as we talk about overhauling our economic systems, climate change and environmental issues suddenly tie in a lot—to pretty much all the aspects of human life. Plus you need air, water, and food and you need ecosystems to support those life-supporting resources in order to have human society."

He knows these ecosystems intimately. His farm produces 85 percent of his family's food—with "family" including the workers and apprentices who live here. It produces all of their meat, eggs, and dairy products like milk, butter, cheese, and yogurt. All of their fruits and vegetables. And nuts, too, so plentiful the farm is feeding squirrels like an all-you-can-eat tree buffet.

Along this road to the vegetable garden, there is a whole hillside of olives—the future crop of Oregon in a hotter, drier world—and an exotic orchard of figs and pomegranates. And because the summer nights are cool and the days are hot and long, a rock wall around the trees absorbs the sun by day and releases its heat through the night. You can put your hands on the cracks, Jacob says, feel the warmth wafting off the stones in the night air.

All of this, it is more research project than commercial enterprise, Jacob says. "It speaks to people from an experience and conveys certain values." They aim to teach, to learn, providing for themselves and a few others as they go, a little income from the produce and the animal breeding, his mother a doctor in town.

He slips through a garden gate, a gray cat trailing him. Inside, the cabbage is monstrous and there is a long row of Roma tomatoes for sauce and ketchup and canning for winter. The greens are vigorous enough to look like an invasion, and there are lettuce and artichokes and brussels sprouts. The celery has an issue, he says. It's going to seed too early. He worries that the stalks will get hollow so he samples some, says it's stringy.

In the adjacent greenhouse, the scene turns tropical. Bananas and passion fruit, their greenery twining to the roof, warmed by a wood-fired steampunk radiator. There are avocados and lychee. Longan and waxy peppers and plump Brandywine tomatoes. Jacob warns about the lemongrass, its scissoring blades. Shows me the plantains, the makrut limes, the wet crops submerged in buckets: ginger and turmeric and water chestnuts, leaves unfurling like tiny lily pads. More citrus stands nearby, bright lemons and oranges and grapefruit. There are edible cacti and aloe vera in tall, thick stalks for use as a salve on the cows and for burns.

For a while it is just quiet while we eat things.

"Science doesn't tell you whether you should save the planet or not," he says later. "They're just telling us what *is*. Apparently the human race seems to be hardwired to go in this direction where we just destroy everything."

Outside, when his father asks a question and there is a quick conversation in their native French, Jacob turns back to me with a lisp on his tongue while he works his speech back into English. He says they are experimenting with buildings next. Making rock-wall foundations with self-healing lime cement and building with the farm's timber, structures to last five or six hundred years, varnish and pine tar to preserve the wood.

This timber is much of what's at risk here as the climate breaks down. Three-quarters of the land on the farm is held in a nature preserve, and the effect is that all around its operations—around the garden and the orchards and the animal pens—towering Douglas-firs cover hills that roll away in all directions. They stand around us now incredibly vulnerable, under assault on several fronts: by increasing wildfire, by drought, and by corporate interests that have set sights on bringing a liquefied natural gas pipeline through nearby land, increasing the wildfire risk.

This area of Oregon is known for its timber. There are nearly 2 million acres of commercial forestland in the surrounding county, Douglas County, including some of the oldest growth stands in the world. And even while the cutting has been curtailed since the heyday of Oregon logging—in famous standoffs with environmental groups over spotted owls—there's enough timber left that at least a quarter of the people living here still make their living in forestry products.

Part of building the farm as Jacob's family envisioned it—as a demonstration of how a single landscape can thrive off itself—is to build from the wood that grows here. Hence the sawmill, the logs milled there. Over a hill or two, on some part of the farm I have not seen, Jacob says there is a geodesic dome in some stage of reconstruction, like Buckminster Fuller might build if Buckminster Fuller lived in an Ewok village.

The timber is among the reasons for Jacob's involvement in the case. Since 1970, the length of the wildfire season has grown by 78 days in this region, from 5 to 7 months. Heavily wooded parts of Oregon like this are among those that burn the most. In 2017, there were 160 days in which the smoke was so thick it was considered a health hazard. Which means that summer workdays on his idyllic farm now require kerchiefs and sometimes respirators. Jacob says the overall increasing heat makes it harder to work. And he sees how the trees and the plants and animals all fare worse in extreme conditions. There is less snow in the winter, too, which only adds to why patches of timber are dying from drought stress, made more susceptible to insects and to fire.

All of this is, of course, textbook climate breakdown. Hotter, drier weather makes for drier soil and thirsty trees that burn fast. It also increases evaporation, drying out plants already stressed by drought and lack of snow and by the pests that make trees more vulnerable to flame. Lower humidity also allows fires to spread. Studies across the West have found global warming has lengthened fire seasons and worsened the impacts of fires in several states. And in one analysis by the Associated Press, the amount of acreage destroyed by fire across the West has more than doubled over the thirty-five years the world has warmed.

For the past eleven years, a liquefied natural gas (LNG) pipeline has been proposed to cross the property that abuts this farm, traversing most of southern Oregon. On innocuous, color-coded maps, this investors' dream pipe winds its way from the drier climes of northern Nevada, the end of the Ruby Pipeline, to a planned export terminal on the west coast of Oregon. In between, it spans some of the most fertile farmland in the state, most of it privately held, plus 70 miles of public trees and managed grassland and two reservations: the sovereign lands of the Coquille Tribe and the Klamath Tribe, only one of which, the Coquille, is amenable to the crossing. The worry of a related explosion feeding yet more flame runs deep here. Any outbreak of fire could wipe out not only the timber but also the farms. On this farm, the plants, the animals, the houses and sawmill, the greenhouse are all at risk. It is an issue that transcends the usual partisan politics among Jacob's neighbors. And it is as much about property rights as it is all the rest. In the summer of 2015, Jacob's neighbors, regardless of political affiliation, were organizing fiercely. Then came this text:

ALEX: Hello, Jacob. My name is Alex and I'm a mutual friend of [name]. [She] tells me that you are involved in the LNG pipeline debate. I live on my family's Century Farm along the Umpqua river near Elkton, and I will soon be suing the government regarding the pipeline. Let me know if you would like to get involved with the lawsuit or if there are any other ways for us to work together on this issue. . . .

JACOB: Wooo!!! Thanks so much for contacting me, mate! . . . I'm currently working to organize an event in September, possibly to coincide with the Hike the Pipe people arriving in our neck of the woods. We are building pressure on the commissioners here to commit to do everything in their power to stop this project. I'd love to get in contact with you and see what opportunities there are for us fellow farmers to work together on killing this pipeline. Looking forward to talking with you!

ALEX: The September event sounds like an awesome idea! . . . P.S. Apparently they are already spray-painting trees in the pipeline route in the Umpqua National Forest. I was up there yesterday, and it is quite an ugly sight.

The route between Jacob's farm and the farm of Alex Loznak makes the best sense to a bird or a natural gas tycoon. Between the hawks and the winding two-lane roads, the Douglas-fir and the gravel drives, the hills are bare in places from dry soil and from shade. There's the Umpqua River flanked by cropland and oak savannah and cattle fences. Simple homes with LP siding, the odd lawn, dot their way into the foothills west of the Umpqua National Forest.

If you were to walk deep into these woods to the south, east toward Crater Lake with its sapphire water, over the ridges in between, you would see the towering firs marked with spray paint charting the route of the Pacific Connector—this the benign moniker bestowed upon the 3-foot metal pipe aiming to carve its way through.

In part due to protestations from farmers along the route, including young farmers like Alex and Jacob, the plan was snuffed out by the Federal Energy Regulatory Commission in the Obama days in 2016, but it was revived in 2017. That summer, following assurances from the Trump administration that approval of the pipeline was imminent, the trees along the route had been spray-painted with Cs for *cut*. They included old-growth timber along the Pacific Crest Trail, the brutal north-south hiking route made famous by Cheryl Strayed's *Wild,* which inconveniently bisected the pipeline's route. It was as if democracy had been canceled. Process a thing of the past.

"These are, like, ancient. You know, five-hundred-year-old trees," Alex says. Bearded in a gray T-shirt, brown hair combed forward, he is standing on a wooden ladder between the Umpqua River and his family's farm, ready to help his father unblock an irrigation pipe so they can water the scorched hazelnut trees. The pipeline hasn't gotten all of the approvals it needs to go through southern Oregon yet, but the Pembina Pipeline Corporation that proposed it is preemptively "already anticipating it's going to go through and spray-painting," Alex says.

Gary Cohn, economic advisor to Trump from January 2017 to April 2018, made approving the pipeline a high priority. A former Goldman Sachs COO, he had just helped the government of Greece conceal billions in debt from overseers in Brussels by pitching officials there on products akin to payday loans, then punting the inevitable economic crash as damages ballooned. He'd only recently resigned from Goldman Sachs to accept a post advising the US government on its own economic strategy, taking with him a $285 million parachute and perhaps wanting accolades for accepting a salary of only $30,000 a year from the public. As a government appointee, Cohn had four primary goals: infrastructure, the gutting of Obamacare, financial deregulation, and a massive tax cut to one-percenters. He had pushed to authorize the Pacific Connector as part of infrastructure plans, no doubt egged on by investors who spied fantastic returns in funding the march of what many Oregonians viewed as an albatross across the state to export energy abroad.

"It's a real shame," Alex says. "The executives of the company met with Cohn," he says, referring to a meeting Cohn had with Don Althoff, chief executive of Veresen, Inc., the project's original owner. Afterward Althoff granted an interview with Bloomberg News detailing the government's support of the project. In the news article, Althoff "was ecstatic. He was saying he's so happy we have an ally in the White House now that's going to push through the pipeline."

Meanwhile, fossil fuels are a hard sell on Alex's farm. By the time I am there in August 2018, there have been five solid years of drought, and several of them have been record setters for heat too. It's how Alex proved his own standing in the *Juliana* case.

The seven-generation farm is 500 acres, about 75 acres of that in lowland and the rest in a mix of Douglas-fir and oak savannah used for cattle grazing. The main cash crop of the lowland used to be prunes, back when Americans had more of a taste for them. But now only a bit of that legacy crop remains—plum trees whose output makes for small sales—and much of the lowland has been converted to hazelnuts. Alex's father, Robin, is waiting to hear about organic certification and in the interim has put a bunch of bird boxes in a barn and let it fill with several generations of appreciative owls. Forfeiting the barn as a posh owl kingdom was a fair trade for the owls' help in gobbling up pests, he says.

Still, the hazelnut saplings are being killed off by the heat and face the prospect of a water table that may fall year after year. Now the river—the farm uses the Umpqua for irrigation—is so low you can throw a stick across it in August. This is where Alex and his father are wrangling their blocked irrigation pipe, in this trace of a river. The bottom is not only water but exposed mounds of stone with tiny fish swimming in the bowls and rivulets carved in their crowns. After the lawsuit began and the government asked Alex to prove he had been harmed by climate change, he turned up roughly fifty pages of records, many of them documenting the cost of replanting hazelnut trees when they die. The rest was asthma attacks. You can't put a record to fear, though, or the concern Alex described to me when earlier in the day he pointed from a kitchen window at a ridge that flamed in the nearest wildfire in 2016, so close it rained embers onto the farm.

Throughout the day, wandering from the kitchen to the river, to the lawn between the roosters and the prunes, in front of the simple house Alex spent half his childhood in, next to the house that his grandmother once lived in, and on a tour of the hazelnuts and the irrigation system, I learn about Alex's role in this case. It is not just the role of rural farm kid—not anymore—or as wingman to Jacob in representing Oregon's conservative south. It is also, and perhaps most interestingly, ambassador to the Ivy League.

When Alex explains this to me, tells me about his day-to-day life at college, it opens a window onto exactly how outside the box *Juliana* really is. This idea of the courts telling the government what to do—it offends legal minds even in liberal institutions, where climate science is a given but the rules of law are rigid as ever.

Before the day Alex heard Mary Christina Wood—the professor who developed the legal theory underpinning *Juliana*—speak to a local climate coalition, arguing that climate breakdown was a constitutional issue for the young, and before he heard about the *Juliana* case, too, and called up Our Children's Trust and signed on, Alex was a kid who'd set his sights on going to an Ivy League school. He'd been pushed around enough in public school that he wanted to be an example for all the other local kids who were dying to get out and do something big.

The case was filed on a Wednesday in August 2015. The following Saturday, Alex left for Columbia University.

Within a few days, he went from his farm on the Umpqua River, from which Eugene seemed like a metropolis and a restaurant was a rare indulgence, to living in New York City and being a plaintiff in what was about to become one of the most important lawsuits in the nation. There were classes and lectures and pleadings and hearings, the travel back and forth to accommodate it all. Alex was up to the rigor, but the fact of the day having only twenty-four hours was problematic at times. He was beyond busy. No time for carefree. And to say there was culture shock in this transition was an understatement. In New York, he was suddenly face-to-face with things he'd had no idea about. Things "like building architecture," he says. Or other students who'd gone to prep schools. He remembered what the timber baron who interviewed him for Yale had said: that the Ivy League wanted him round like a bowling ball. Smooth. He was a stone in rough water.

In those early days, Alex was surrounded by some of the greatest legal minds and environmental thinkers of our time, many of whom took an interest in the *Juliana* case. Some had an interest already. Like

James Hansen, the former NASA scientist who had told Congress that climate change was urgent and man-made back in the eighties. Hansen was a professor at Columbia and an expert witness for the plaintiffs. He had also joined the litigation as a plaintiff to represent future generations—the only plaintiff in the case older than nineteen when it was filed. The Nobel Prize–winning economist Joseph Stiglitz, who estimated the economic cost of not responding to climate change, was also an expert witness for the plaintiffs and a professor at Columbia. In fact, much of the climate research that laid the framework for the plaintiffs' claims was a product of Columbia's observatory, which had been accumulating earth science data since 1949. That the facilities and the faculty at Columbia were enough of a lure that students interested in these matters gravitated to the school made *Juliana* all the more relevant to Columbia students. The legal community was similarly tuned in, Columbia being home to the Sabin Center for Climate Change Law, which tracked climate litigation worldwide. Law professors could not resist debating whether Wood's constitutional arguments could hold up in court.

This legal skepticism stemmed not from the merits of a case for the climate but from the fact that *Juliana* tested just how much the courts could tell the executive branch of the government what to do about it. Law professor Michael Gerrard, an outspoken doubter of whether the *Juliana* approach could survive review by the courts, even debated Alex on the merits of the case—something of a mock trial—in his class.

As much as Gerrard and other legal scholars supported the cause—Gerrard was the founder and director of the Sabin Center—he was not unlike other legal purists who could not imagine that a court would assume the authority to act on climate change. He and others thought the government's argument that only Congress could make climate policy and the president had to carry it out or make his own would probably prevail, even if the government's promotion of a fossil fuel energy system had a hand in accelerating global warming.

"The biggest issues are the legal issues. Is there a constitutional right involved here? Does the public trust doctrine apply? Do the federal courts have this kind of power? Those are not factual issues," Gerrard says later, noting that prior court rulings had given the EPA, not the courts, the power to regulate greenhouse gases. "It's not appropriate for the federal courts to be making those decisions."

Alex could be good natured about this kind of skepticism. So much so that after Gerrard asked him during class whether he planned to put the president in jail, Alex could only smile. There were a few *Juliana* groupies. Few enough that Alex could still be Alex. But in the weeks before the trial, those days felt numbered. Receding. Like a past he was already nostalgic for.

By the time we are sitting at his kitchen table, Alex subtly battling the summer flies, he is not that insecure freshman anymore, instead a rigorous intellectual, eloquent and bespectacled, about to start his fourth year at Columbia and intent on becoming a lawyer and a permanent fixture in the climate fight.

In addition to homework and hearings and all the rest, he had read everything to do with the case. *Everything*. And he would discuss it with his Columbia cohorts, the scholars of law and science that they were, so much that in the intervening time he had developed an almost detached relationship to the litigation. In the moments that it felt like his, when he carved out time to think about his own stake in the case, it made him anxious. And in the moments when it wasn't his, he still thought about it, but with curiosity. He could not tell which way it would go but had the benefit of hearing a lot of very smart, very engaged people predict the ways. To sit back from it was to protect himself from how it could end, and from the stress of how important it would be to the rest of his life, to the rest of all of our lives, no matter how it turned out.

"Even if we don't win, I think that this case will really lay out a vision for what the next several decades can and should look like," Alex says. "And I would even go so far as to say that that is really my generation's vision for America. That's the country that we want to live in, and we want to help build and create into our middle age."

He tries not to let the tension overwhelm him. In certain moments, he can only defer to his grandfather, an attorney in Roseburg, Oregon, who when speaking of the law and what to expect from a trial was fond of saying, "It's like a pig on ice."

In other words, you do not know which way it will go.

JACOB: Yeah I'd be interested in being a plaintiff. . . .

ALEX: How old are you?

JACOB: . . . 18

ALEX: Ok, so the lawsuit is being done by Our Children's Trust, which is a group out of Eugene that orchestrates climate change lawsuits on behalf of youth. The aspect that they are suing over is the carbon emissions from the pipeline, and they are representing plaintiffs under 21 yrs old from Oregon.

JACOB: . . . Yeah, definitely would be willing to be a part of that.

The trial would not be a small contest. But by August 2018, the success of *Juliana* was quieting, or at least surprising, many of the skeptics who'd first seen it as a longshot. The plaintiffs had proven their stake in the climate fight in court. Plus, the government's attempts to dismiss their case from the lower court had all failed. And Judge Ann Aiken's ruling that Americans have a constitutional right to a stable climate was significant, a first of its kind.

Still, what lay ahead was not just the upcoming court battle but also a battle for public perception, the right to hold the moral authority that the case conveyed.

"Sometimes I get a feel that, for the nation we're suing, since this is about the US, there should be way more interest from US media," Jacob says. We are still walking around the farm, Jacob explaining everything from the mill to the fruit orchards and gardens, with asides about life as a federal litigant and an up-and-coming media darling.

As a Canadian immigrant with French as his native tongue, Jacob has become a fan favorite of the Canadian and French press, both of which have taken an interest in the case. And in that way, the media approach to *Juliana v. United States* is typical of the media approach to climate change: the foreign press wants to know what the United States is going to do about it, and the mainstream press is only cautiously attentive, allergic as it is to the attending histrionics and the runaway comment threads. Still, the world is watching, even if many Americans are not. After all, until Obama pushed US involvement in the Paris Agreement—the nonbinding treaty by which world nations would work to lower carbon in the atmosphere—the United States had been the biggest obstacle to United Nations negotiations over climate change. America was one of the few countries that had failed to ratify the Kyoto Protocol, a prior effort. And the US Senate had approved a resolution disallowing the ratification of a binding climate treaty in 1997, even though the United States is responsible for one of the largest shares of carbon dioxide emissions in the world and is among the chief consumers of manufactured goods that boost emissions in nations like China.

Jacob is game for the interviews. But the attention makes him nervous. He is not atypical in this way. As the trial looms large, anxiety is a pretrial bedfellow. The plaintiffs aren't just buying plane tickets, they're making contingency plans for life; taking time off school, off work, off friends. The natural outgrowth of all that excitement is a lot of stress. They have all begun to worry about things, even though the things they worry about are not the same things.

For Jacob, what troubles him most is that he will soon be called upon to stand in front of the litigation while the rest of the world stares at him. To say it is unnerving is insufficient. He's been trying to quiet the part of his brain that is telling him he could be made pretty famous by all of this, because he is not that kind of person. He is not someone who has aspired, ever, to stand in front of a crowd. He's a creative guy. Writes poetry and reads a lot and likes to play Irish music. He doesn't consider himself an activist, didn't grow up in the movement.

The facts just happened as they did. He started protesting the pipeline, then spent five days fasting in front of the Federal Energy Regulatory Commission in Washington, DC, to underscore his point. He joined the lawsuit, went to Standing Rock, attended a World Peace and Prayer Day ceremony, not necessarily in that order, and meanwhile the lawsuit grew legs. In time, court after court decided in the plaintiffs' favor, and the lawyers from Our Children's Trust proved themselves capable of leading a case of national—international—import. Now here he is. Center stage, imminently facing the curtain call. And while Jacob has thought deeply about the issues he chose to litigate, and believes with every bit of himself that this case is important, those things matter little to the question of whether he sees himself as a leader in the vein. He does not. And it doesn't matter.

"It's not really that I want to . . . give an example. It's more how do I live up to that responsibility to set an example? Because I'm here. People are looking at us," he says.

So in the summer of 2018, he had been spending a lot of his days thinking about how to live up to being a person that a lot of other people were about to be looking at. And the best way he could imagine standing in the spotlight was to simply do the right thing. He was studious, so he studied what that meant. And in observing how others had done it, he had adopted an approach that was two parts Martin Luther King, Jr., one part Samuel Beckett. This happened because he was watching the Parkland youth closely. And in an interview on *The Daily Show*, some of the youth who had taken on the gun lobby after seventeen people were killed by an active shooter at their high school talked about the teachings of MLK. Jacob realized he could use these teachings, too, particularly the fourth: the notion that when you act on behalf of a cause, it is not about you. Whether someone throws a bucket over your head, or whether you are jeered at in the back of the bus, when you are speaking for a cause, to do so with grace and nonviolence is an art. That when the world is watching, to stand strong peacefully is to speak a language that people can hear across the political spectrum. To be gentle is to be understood rather than to create opportunities for hate or division or for anger.

The anger, it is out there.

"I try not to think too much about things I can't control. I'm not going to leave this lawsuit and I'm not going to tone down my public persona or tone down interviews or retreat. So then the question becomes, What is the best way to . . . present the message? I believe that if we can do it in the right way, it almost undercuts . . . the backlash."

He is preparing for it anyway. Thinking about cybersecurity. Thinking about his personal security. He wants to unite people. Reach past the hostile divide that political and environmental discourse has become. And to judge how well he is doing, he need only walk to his nearest neighbor's house—Douglas County is awash in conservatism—and try on the things he might say. Try to listen. Try to see if he can meet people around the values they all share. Like keeping the pipeline off their farms. Protecting their trees and their animals.

But this is dicey terrain. Before his family purchased the land, there were stashes of guns in the hills and they knew the area was a stronghold for white supremacists and other people disinclined to share his views about most things. Yet here he is, long hair and flannel, as much a part of this place as anyone else. His tolerant Quaker disposition and his farm experiment in zero-impact living are not typical. But solar? They can all agree on that. He is trying to find the lines, reading a lot about how to talk to people through common values, set aside judgment and know that people can be reached. That if you find them where they live and let them really know you, they will listen.

One neighbor "hates wolves and he would kill all the wolves in Oregon if he could, so he's full of contradictions. . . . He'll be saying something one minute and be like saying completely somewhat the opposite thing politically the other minute. And I think a lot of people are like that."

Jacob shrugs. We are walking between vegetables, an ATV having just passed from the other side of the greenhouse. He says he doesn't feel bound by the political reality of what is happening in America. He only feels it is right to carry on.

"We're giving it everything we can. It's how a lot of these civil rights leaders throughout history gave. And you have to have that outlook," he says. "If you're doing the right thing and you're going in the right direction, every step is a victory. You can't lose." You can only not do enough, or lose touch, or become too attached to one part of what you're doing and not see the bigger picture while it crushes you.

He knows he will fail. Sometimes. And he isn't naïve enough to think that a lawsuit can change the world by itself, or even that humans might not be wired to just end up in one of those megalopolis planets like in *Star Wars*. Maybe that's just how evolution goes. Maybe there are other planets.

But if the moment is such that Pope Francis and a lot of Indigenous leaders are all telling people the same thing at the same time, and scientists, too, and he is a person who is chosen to stand in front of that, then he will try to find the thing he can do with dignity. Maybe it will catalyze the next thing. Or at the very least, maybe it can communicate a certain value. He hopes his values are obvious every time he stands in front of his farm.

JACOB: You know [name]? He owns the property above ours. The pipeline crosses on his property twice. We're not in the route I believe.

ALEX: Yes I do. He's legit.

JACOB: He is! It's great to have some Republicans on board too, this fight transcends all political parties.

ALEX: Yes, there are so many angles, property rights, climate change. . . .

The trial is in eight and a half weeks, so depositions are happening at a furious pace. The government delayed so long—fighting requests for records, refusing to depose witnesses—that the attorneys on both sides had fifty days to complete forty-eight depositions.

This day on his farm, Alex has just had the "dubious honor" of being the first plaintiff deposed. Not just in his case. In any atmospheric trust lawsuit ever. So the playbook was his to invent. The government's attorney was a good one, he says. And it was hard to know what the strategy really was as she grilled him. She asked him questions about causation, mostly. How did the government cause wildfires? How did the government cause ocean acidification? Questions that were best answered by experts and would probably be objected to in court. Whether she was looking for slip-ups or signs that he'd been coached or something else was unclear.

"I was stubborn in terms of giving answers to the questions. But eventually you have to give answers to the questions," he says. We are on the lawn in front of his house, the too-many roosters hatched from the henhouse pecking at the grass. He thought they'd been harder on him than on the younger plaintiffs.

He sat in on some of those depositions, too, proud as he watched. Alex felt at home with these people in a way he did not with others. When his classmates talked about tuition at their prep schools—$20,000, $50,000 a year—he felt the acute gap of being homeschooled on a farm, of attending public high school in Roseburg, of having arrived at the Ivy League as a fish out of water. But then he'd often been that person, the homeschooled kid who didn't fit well with the other kids who had been classmates forever. But here? With these people? They were his and he was theirs. Across age and geography and race and economics they fit, each their own important part of a thing that was becoming like a child itself, their child, growing up, leaving the nest, something they often watched with the detached wonder and fear of worried parents. He marveled at the young plaintiffs, how much farther ahead they were than him at their age. He thought Avery McRae could grow up and be president. That all the young plaintiffs could speak unimaginably well. And Aji Piper—who in his deposition, when asked the cause of ocean acidification, had pulled the equation for it out of his head, explaining to a baffled attorney that this was basic high school chemistry—Aji made him laugh.

In the background as he speaks are the old plum trees, the family dogs mingling with the roosters.

As the depositions continue, the trial a mere two months away, Alex has the sense that while many people are aware of the lawsuit, there are many more who are not aware that it is more than a publicity stunt. He wonders all the time what the reaction will be when it becomes clear that the case is legitimate. To those watching, it seems obvious the case will win at the district level—it is about to be heard by the same judge who has just ruled that Americans have a constitutional right to a stable climate. And while nobody knew what would happen at the appellate level, a win in the district court would signal what America needed to know: the shit is serious.

When the Supreme Court rejected an emergency petition from the government to toss the *Juliana* case in July, calling it "premature," the decision noted that there were "substantial grounds for difference of opinion" and that the US District Court of Oregon should take them into account while discovery was still under way. Justice Anthony Kennedy took the lead on motions arising from the Ninth Circuit, and this decision was one of his last before retiring, giving Trump a second opportunity to nominate a Supreme Court justice.

That this petition had landed right after Kennedy announced his retirement was not escaping notice. Trump had already nominated Brett Kavanaugh to succeed Kennedy, who'd often been a swing vote on climate issues. Kavanaugh, by contrast, was a DC appellate court judge whose past decisions limited agencies' ability to regulate greenhouse gases without guidance from Congress. Though confirmation hearings are still ahead, his appointment could have a huge impact on future decisions by the court, including future petitions to dismiss *Juliana*.

The nation has yet to learn of the sex assault allegations against Kavanaugh, and those confirmation hearings have yet to become a kind of de facto trial for Kavanaugh's alleged juvenile attack on a coed. But it is already going to be a close vote. The dissent of just two Senate Republicans will halt the appointment. Meanwhile, Senator John McCain's death the day before my meeting with Alex seems to signal an end to a certain centrist era. The whole nation is on edge, but Alex is not willing to speculate on what effect these changes in the composition of the Supreme Court could have.

Elbows on knees, cargo shorts on his farmhouse lawn, he says at its heart, *Juliana* is a constitutional case, not a political one. And like the attorneys involved and the rest of the plaintiffs, he is counting on the courts to set politics aside.

What he predicts instead is that the government will use every tool it has to squash *Juliana* before it goes to trial.

CHAPTER THREE

"Not a Problem Individuals Can Solve"

With six weeks to trial, the depositions still under way, I head to Colorado to meet Nick Venner, the seventeen-year-old asthmatic who is litigating, among other things, the relentless wildfires that make it hard for him to go outside.

I'm like a lot of people when I travel (a lot of green-leaning people, anyway)—overwhelmed by how much garbage I'm responsible for, how many plastic containers, how many straws. The travel involved in reporting this trial hits that home. I haven't left the Denver airport before agonizing over the bottled water I bought there—no water fountain in sight and no refillable bottles on offer, lest we destroy the market for single-use plastic at Denver International. Breakfast comes in disposable cardboard with a plastic fork. And by lunch there is a yogurt container I don't know how to recycle and a plastic wrapper from string cheese.

My rental car—a midsize SUV—also layers on the guilt, guzzling gasoline while I drive to Nick's hometown and to some of the wilderness Nick aims to defend. By day two, even the egg whites at Starbucks come in a cardboard container. Another day, another plastic fork. I think: I'm doing this work for a reason, but I'm doing it wrong.

So it's a relief when I finally meet Nick, because he assuages my consumer guilt in a few words: "I'm kind of of the belief that one person shouldn't have to worry about this. This is not a problem individuals can solve. It's not caused by ordering straws in restaurants." It's our systems and the related industrial processes that are to blame for climate change, he says. This is what the *Juliana* plaintiffs want to address, and why they're suing the government instead of suing the rest of us. They're not out to change people's personal habits. They're not telling us all to live differently. They're making a case for a better society. One in which the health and survival of our civilization is the thing we prioritize first, not a political second to taxes and school bonds, or an obstacle to religious and political beliefs that align themselves with denial. This means living in systems that foster a stable, safe environment, and not torturing oneself about daily habits.

I like hearing this from a teenager. Like he's letting me off the hook. Which he hates, by the way. Most of the plaintiffs hate this idea: that it's all up to them. As if the rest of us can just sit on our hands.

"Do you know what an axiom is?" Nick asks me. He is a lover of math and science, addicted to chaos theory, and very quickly it is clear how much he can talk about numbers until I do not know what he is talking about.

We're on the backyard patio of his home, a contemporary ranch with a shade roof over a concrete slab. The chairs rock, and when Nick's emphatic, he bounces in them. So does his voice—it has a higher pitch when he is thinking out loud or surprised by a question, which gives the impression of someone who is always upbeat, always ready for what's next. I find it endearing, but it bothers him. Nick's vocal shifts are symptomatic of his Asperger's syndrome, and they make it hard for him to give more public talks about climate change, something he wants to do.

This sitting—it wasn't our original plan. Our original plan was hiking. But when I knocked on the door of Nick's home in Lakewood, a suburb of Denver, he apologized for the weather right away. Which was his way of saying that he didn't want to go hiking anymore. It is roughly 90 degrees outside, and Nick hates the heat. Hates that it persists even though it's September. There used to be more snowstorms than heat waves in Colorado this time of year. Now the heat is unceasing. Last year too. Nick likes to ride his bike—it's 3 miles to school—but says the weather keeps him from doing it. He has exercise-induced asthma made worse by the higher temperatures and the smoke, which accounts for his being, in his own words, "Kind of unfit, typically."

It's been like this in 2018: fire season is still vicious in parts of the West, the air thick and tinged with the smell of burned timber. Even in the cities, there are days when the waning moon hangs in the sky like a pink fingernail, dusk seems to fall at 3 p.m., kids are kept indoors, and normal chitchat takes the form of complaining about the air. Colorado is getting the nastiest of it, having the worst fire season in the state's history, with more than fifteen hundred fires this summer. This is life in the West now. Longer, hotter, drier seasons; more flame. And for Nick, that spells a lot of hiding indoors, his mother's air purifiers whirring on the second floor.

He says the smoke here has been awful. Dense and heavy. And that even though he thinks the benefits of going outdoors vastly outweigh the benefits of avoiding air pollution, he has been unable to cope with it because of his asthma. He's had trouble breathing, and even when he braves it, he can't move as fast as he otherwise does, and as a result can't calculate how long things take and whether they are safe for him to do.

"It's been really bad," he says. His mother, who is similarly afflicted, has hardly left the house. Now, Nick just wants it to be autumn so he can ride his bike to school, hike whenever he wants, sit outdoors and not feel like the sun is slowly roasting him.

I know what he means. A day earlier, when I first arrived, I took a short tour of Lakewood—the seven-lane roads and the strip malls sprouting alongside housing developments—and stopped at Sloan's Lake, a park near Nick's home. The grassy meadow looked dry enough to ignite with a hot breath. There was a light breeze, but it was arid and warm and offered no real relief from a sun that scorched the skin in minutes. Though the lake still held water, trees were standing dead in the meadow in clumps of dry, twisted bark, giving the strange impression they could simply combust. Looking around then, from the lake to the nearby Rockies and the timber-covered foothills, it was easy to see how the most innocuous flint could set the landscape ablaze. Even the soil was parched and dusty. I could imagine it whipping an untamed flame onto wind.

Until he can vote, Nick says it helps to subscribe to chaos theory. The idea that a butterfly can flap its wings and cause a typhoon somewhere else in the world, maybe years later. After all, there are other ways to make a difference in climate change. The idea that one's actions can have ripple effects and that you can't really know what it is you will achieve, only how to act with intent, there is power in that. Whether suing the federal government, talking to that one person who becomes a politician later, becoming the scientist who writes the reports that policy makers then read, or talking to a reporter who doesn't know what an axiom is.

I ask him to explain. Nick says axioms are the bedrock of mathematical systems. "If you're coming up with, like, a mathematical construct of things, what it essentially means is that axioms are the core, the fundamental beliefs and values."

This is the problem with the way in which a lot of people relate to climate change, he says: they subscribe to a fundamental belief that is not as sturdy as it seems. "Humans developed scientifically in a time where the climate was unusually stable," Nick says. "We've built a human civilization on the fact that, well, grain will grow on the same field year after year, that the sea will not rise and flood the homes, that the trees will still continue to grow, that the weather patterns will continue to remain the same, and temperature will remain the same. It's an underlying axiom of sorts.... That's why we build farms and we cultivate the land. We assume that each year is the same as the last. Each year will bring the same amount of rain at the same time and it all will be predictable. That isn't really the case."

His cockatiel, Manu, is perched on the table in a modified cat carrier, Nick having tried a smaller carrier moments ago. The bird is gray with a yellow head and bright orange circles on its cheeks. Nick thinks the color combination is called pearl, if you want to know. And throughout our talk, as he did just moments ago, Nick defers to the bird. He can tell by its tweets and warbles, by the angle of its crest, by the way it swings its body side to side, whether it is agitated. He takes several breaks to acknowledge this avian commentary: loud chirps over the sound of a lawnmower, a stiffened crest—possible predator? Nick turns in his seat.

While a lot of people think it's hubris to assume a human can change the planet, he says, humans, together, have plenty of ability to change ecosystems, and they have. "It isn't really one single person's actions that make that happen. It's that by living in a system, you're essentially contributing to the entire thing." That a person's actions become inconsequential if you break them down small enough "is actually a reason why I've always tried to avoid saying the key to fixing climate change is to not get straws at restaurants and recycle more." His family has done all it can. They have EVs, solar panels. But they're still contributing, too, Nick says. "By living in a system, you are perpetuating that system. We still get electricity. We still get water." Thus his line that climate change is not caused by people ordering straws in restaurants. "It's caused by systems. It's caused by the industrial processes which make those straws. It's caused by the power generation that makes these lights." This is what he most wants people to recognize: that wholesale success means shifting focus from personal habits to the more destructive collective habits we all share.

He rocks back in the chair. He is wearing one of the button-down patterned shirts he favors, a nod toward Hawaiian but with a more professorial flare. "I'm a kid," he says. "All I can really do is influence my parents' financial decisions and educate other people so they will take on action for me."

This is how Nick has gotten so good at haranguing his parents, has almost always been good at it. His mother, Marie, tells me Nick has been obsessed with environmental causes for so long that she can remember noticing it when he was still small enough to ride in a booster seat. One day she was listening to a lecture in the car by Richard Rohr, the ecumenical teacher and environmentalist, when Nick piped up, to her astonishment, and declared from the backseat: "This really makes sense!"

Ever since, he has been so unrelenting in his pursuit of activist aims, he once lectured his parents all the way from Aurora to Lakewood over their choice to buy a gas-powered vehicle instead of an electric one. He was so offended by this impending purchase, his critique so unceasing, that his parents ultimately scuttled the plan and sold their second car to afford an EV. Now they have two EVs. And the solar panels. And in between Nick translated *Laudato si'*, Pope Francis's encyclical on the care of the environment, from Italian with Google's help because he couldn't wait for the English translation. He also made climate presentations at a school full of doubters, often at the expense of social ease. And by the time I meet him on the back porch, he has forfeited a lot of social acceptance, plus his former church congregation of climate skeptics and vast territory in his brain for climate justice. It is not so much that they raised him this way, Marie says. It is more like they couldn't stop him.

Nick says he came to understand the more specific threats associated with climate change when he first began learning about it in 2008, when pine beetle damage was at an apex in Colorado forests. He remembers the month because Obama had just been elected. At the time, Nick belonged to an environmental club, but when he tried to introduce climate-related issues, "they wouldn't have it," he says. The era was such that "people could deny climate change and you kind of had to respect that as a difference of opinion."

But when he first took stock of pine beetle destruction, "what was kind of immediately striking was I'd seen these impacts before. Like, on this one route, I've seen these dead trees before and I just never thought what the cause would be."

Nick's pleadings in the *Juliana* case include the charge that pine beetle damage, along with wildfires, has forced him to stop visiting some of his favorite places. Now so many years later, the pine beetle problem is worse. Just how much worse is best witnessed by traipsing outside, something I decide to do without an asthmatic kid in tow at the peak of the fire season.

Rocky Mountain National Park is one of the places hard hit by pine beetles. On the day I visit, the sign at the entrance gauges fire danger at *very, very high*. It is September 18 and the aspens are finding their fall yellow, a few leaves drifting on the breeze. This scene is every bit as picturesque as any brochure would portend, save for the drought-stricken grass and this quest to take in the views of beetle wreckage, which are supposedly optimal from the main road.

Earlier in the day, when I asked for tips on where to look, a ranger kindly offered intel. From behind a counter, she proffered a visitors' map, then pointed to a spot called Timber Creek. It got its name from the massive lodgepole pines that used to stand there, campsites nestled underneath. But when pine beetles claimed the trees, and the dead husks loomed over campers like a catastrophe waiting for a breeze, the park service cut them down. A massacre of another kind. Ever since, the rangers have had a new name for the campground, an inside joke: Timber*less* Creek.

The route to this capstone ought to say it all. The ranger marked the map for the particularly ugly views and also suggested a trail along the Colorado River to get up close with the beetles. Now, I need only drive the one-way route through the park to understand the parasitic bugs that have overtaken some of its parts.

Finding this carnage is actually kind of hard. Not because it's difficult to spot the barren vistas of needleless trees, their spindly branches outstretched like the sad wings of featherless birds. Those are right there. But anyone who has spent time in the West has unwittingly seen so much pine beetle damage that, as Nick says, it fades with familiarity like the details of a wallpaper print. In the Coast Range, in the Cascades, the Siskiyous—it's easy to mistake these naked trees as singed by wildfire, damaged by floodwaters or by drought. In other words, for the things we already understand. But to look again, knowing. The devastation is stupefying.

The first signs are just a few solitary dead trees—lodgepole pines—as the road starts to climb toward Sundance Mountain. Then there's an isolated clump interspersed with swaths of otherwise green trees, gray smudges on a forested hill. Deeper into the park, as the pines rise up all around, I see a bare tip here, another there, this touch of gray against green. The damage seems minor. But then around a bend, suddenly the gray overtakes the vista. The whole of the hillside descends into the valley like a moonscape—rock and dusty soil—with a cover of zombie trees. The trees are still standing, still bearing their needleless branches, but they're dead, done for, over.

Look closer, and the bark is dark, much darker than the bark of live pines, save for being covered in the bits of goo that from a distance look like natural sap. It's not, not entirely. Instead it's tree pitch, the stuff trees push out through their own bark as a defense against the beetles that bore through. The temperature is 74 degrees Fahrenheit outside, hotter at the lower elevations. But even at 12,090 feet, far above the Continental Divide, where the wind is relentless and the trees don't grow, the temperature is still 66, about 20 degrees higher than past averages. The relative warmth, of course, is part of the problem.

It used to be cold here. Very cold. And year after year as winter set its teeth into this wilderness, the penetrating cold and the snow would kill the beetles, keep their populations down and their infestations seasonal. But it takes temperatures colder than –35 degrees Fahrenheit to kill a bark beetle. And temperatures that cold haven't befallen these wilds since the mid-1980s. As a result, bark beetles live and live and live. They infect a tree and breed, and those beetles live, too, and generation after generation, they infect tree after tree. Between 1996 and 2013, mountain pine beetles infested 3.4 million acres of ponderosa and lodgepole forests in the state of Colorado alone, or about 81 percent of the state's such forests. Lately, a fast-growing spruce beetle problem has killed—in just a few years—about half the number of trees that pine beetles killed over decades.

The frequent droughts don't help. Dry, weakened trees can't fight the beetle invasions off. And when fire strikes, those infected trees burn hotter. Of course, bark beetles are native to these forests, and they do have a job: cleaning up lightning-struck and diseased trees, helping them fall and regenerate soil for the next iteration of growth. But when the beetles are so rampant that they start to attack healthy trees, and when even the fire and the insect-loving birds that co-evolved with them can't keep them down, that's trouble. Forests all across the West are at risk for this fate now, and more than 100 million acres from British Columbia to New Mexico are already infected.

Dan West, the state of Colorado's bug man, says this is all bad news only through an anthropocentric lens. In other words, things are going great by the standards of the bark beetle. They like big round trees, and there are plenty of them. And if you consider that it is baked into their species to leave something for their grandchildren, it is hard to argue that we humans have the higher evolutionary ground. The beetles leave trees untouched if under 4 or 5 inches in diameter—they're after the sugary phloem beneath, and the little trees have fewer carbs. Thus, the forests will regenerate and regenerate no matter how many times the beetles infest them. Good news: new forest. Bad news: these will not be the forests we humans hoped to preserve when we set these lands aside as a national park.

For Nick, whether these forests regenerate into something of human benefit later is only one part of the story. In between the brutal conditions of today and that future, nature does not simply cope. There is wildfire, species extinction, tons of ecological damage. "If it's possible to preserve the forests that we have now," Nick wonders, "why would we put ourselves through—for lack of a better word—the *pain* of going through that?"

I feel him. Even though I am like most people in that the parts of this particular wild that I can see come with pit toilets, this mess is gutting. These western forests are the ones that swallow you in their quiet, in canyons carved by glaciers, on mountains where time is longer than a human life, trees towering above the ferns and moss that thrive at the feet of giants. To think that they are all at risk now for a bunch of bugs gone rogue in the heat—the thought of it is almost too big for my head.

That is the problem with all of this, isn't it? Climate change? It is too big for our heads.

On the Colorado River Trail, one doesn't have to travel far to see this devastation up close. Right away there is a little stand of trees with bulbous, gooey drips covering the trunk of a pine. There are teeny holes throughout this goo; exit holes for beetles that have done their damage and moved on, leaving bits of sawdust on the forest floor, bits of tree that have been burrowed out from the inside. These holes look like portals to a never-ending bug kingdom. And they are. The farther I walk, the more there is to see.

Lodgepoles again, and ponderosas too. Many of the trees infected. But because the branches seem to die from the ground up, and the lodgepoles are often bare trunked near the ground anyway, the situation, dire as it is from the road, looks worse up close. Once inside it, walking among the wreckage, I see that far more trees are covered in pine beetle damage than is obvious at first, some fighting so hard to stay alive they look for all the world like they've been dipped in vats of glue. Pitch-covered bark is everywhere. Yellow glue, orange glue. Sometimes it covers a tree, and sometimes it just drips down a side like wax down a candlestick. Some of the trees are valiantly hanging on, with just a few last sprigs of needled branches at the crown. Others are only beginning to die, the low branches turning from needle-covered limbs into dry, curling brambles. The dead ponderosas look from afar like oversized kitchen brushes, something to clean a jar with at a sink. In the vista are ponderosas with red needles—signs of a first-year infection. By next year, those needles will fall too. The husks of their trunks have faded to red, the lodgepoles taking on a gray, burned-looking cast.

Almost everywhere else, these trees seem to be shrinking, falling, making way for other trees that can live. On the route out of the park, a hillside of denuded trees climbs up toward a ridge. At the top, stark across a backdrop of green, a row of dead trees stands at stiff attention like somber clothespins. At their feet are more where others fell. Some have tipped into their neighbors and not yet struck ground. They are not very old, not very large, just lying in a mass of useless timber.

The exit is by way of Timber Creek, Timber*less* Creek, where the bewildered campers shuffle among tents in the heat of the sun. The tops of their vehicles make for the high points in this landscape, replacement pines only just dotting the space in between like chubby Christmas decor. The roofs of the cars and trucks and RVs run off into the fading light, reflecting it back to the surrounding hills. They are stuck with each other, these campers, with little greenery or bark between, just a few baby trees and picnic tables scattered in the low scrub. They look hot, unsheltered, confused.

—

To see humanity struggle with things like straws instead of this scale of devastation—or the root causes of the devastation, like warming temperatures, lack of snowfall, and drought—is, for Nick, to see humans suffer from a kind of blindness.

We are still on the back porch, listening to the whirs of his neighbor's lawnmower from our safe zone inside the wooden fencing of the yard. A chicken coop stands empty in a corner, and Manu is still perched on the table in his cat carrier.

This blindness, it's a rerun of a type of human struggle that, in Nick's experience, has its worst example in his religion. He knows that fellow Catholics have ignored the sex abuse scandal in the church or dismissed it as unimportant, that they need this denial because it helps them get on with life and still go to church. He understands why it's just easier this way sometimes, even when people have all the proof in the world. "It's a similar reaction to climate change. 'I know it's happening. I know it's a big deal, I should probably be getting more upset but it's easier to not do that,'" he says. This is the piece he wants to change. Not anyone else's habits. He doesn't want to give people guilt. He wants to give them conscience, and the will to push for a society reformed.

I like this example. I like it because it illustrates how a very serious problem can get bigger, and how inaction can combine with denial until a crisis threatens to destroy a thing that once seemed too big, too sanctified to fail. Similar, say, to failing to prepare for the pandemic that will soon be upon us, even though history and science tell us it is coming. The solution is similar too. It lies in the reaction of people, a common faith in a renewed journey forward, and the support of a congregation to do better the next time around. Climate change walks a comparable path.

"Fundamentally, it's a problem, and it isn't any more difficult from any other problem. It isn't unfixable or unstoppable. It's a relatively simple problem: humans are causing the climate to change based on pollutants. The most efficient way to stop that is to stop pollutants. If you start looking at those and the way to address them, you can get bogged down fast," which is why Nick thinks climate change is forever being set aside by political leaders in favor of more populist issues. But in the end, it's just a problem in need of a fix.

Here, Manu chimes in. And Nick is briefly quiet.

He says he knows people have other things to worry about. Work and school and those big, capital-R Responsibilities. "It's very easy in that stew of craziness to forget about some things. And I feel like climate change and a lot of other social justice issues kind of get excluded outside of that things-you-can-actually-care-about circle."

Then he says if people stop thinking about climate change as an impending crisis, as anything other than a problem that needs fixing, the inertia could give way. People could embrace that their individual role can be to push leadership on fixes, prioritize remedies at the ballot box, and support youth like him who know that change will come from reimagining the world around us, not fighting about the one we already live in.

The straw boycott, it's been a headache for him.

"I find straws annoying," he says. And for a few minutes he struggles to say why until he talks about the amazing, stirring speeches he has heard people give about the evil of straws and how the subsequent call to action has only the one predictable response. Here he imitates: "What do you want? What do we do?! *Don't get straws.*" He frowns, "Like at that point in time I go in the car and tear my hair out."

He knows tiny actions like these aren't enough to address the very serious environmental problems affecting the planet. And his summary of the attempt makes me laugh, makes Nick laugh too. The straw boycott, as the latest in environmental fads, *can* be especially grating. It has come on strong this summer and is at a pitch. Suddenly people who came of age with fast food franchising have seized upon the straw with a kind of venom.

On the surface, it is reasonable enough. For those who don't need them—for example, people with disabilities or lying in hospital beds—straws are useless anyway. And the number of plastic straws used daily—500 million, cluttering the oceans, littering beaches worldwide, even found up the nose of a sea turtle—is a sad commentary on humanity's trash. Why not put an end to the ubiquity of a thing people mostly don't need? It's a quandary we can all handle. But that is Nick's critique of the straw boycott too. That it is a distraction from the scale of the crisis we really have with the health of the planet. And the vigorousness with which the straw is being attacked is also a hat-tip to a cultural preference for ease in solving such predicaments, and to the persistence of the belief that environmental dilemmas can be fixed by either buying things or not buying things.

These boycotts, they always seem to go like this. One day straws exist in unwitting culpability along with the rest of our trash, as they have since the 1950s. Then another, it is as if straws have called in a bomb threat on all of us. Suddenly you have to hate the straw, hate the place that serves the straw and the person who serves it to you. And to join in the hate is to declare oneself of a stripe, of a moralist breed, while others remain in league with the straw, want it, expect it even. These are the people to be judged. I can only imagine the waiterly paralysis that attends this: *to straw or not to straw*. Either way is to risk an upbraiding, and to questionable end. No similar hate is levied at other ubiquitous plastics like the plastic bottle or its lid, more prevalent than the straw in beach cleanups worldwide. And there is still no call to oust the shampoo bottle or the kitchen trash bag, the daily plastics we cannot seem to do without.

Yet this is the way in which mob mentality has governed not the environmental movement itself but the mainstream acknowledgment of it. There is less sense to the approach than a kind of collective convenience. And to Nick's point, straws are an easy mark. So, too, are plastic grocery bags and the plastic rings on six-packs. We can live without them and we do, soldiering on in the crusade against things that are easy to hate. We never seem to target the actual enemy. Or choose our enemies with the kind of logic and cooperation that the climate crisis demands. The humans are still in charge, after all, and often guided by emotional impulse and sometimes, it seems, increasingly allergic to reason.

By the way, Nick says, if everyone avoided straws and recycled as much as they could and did their best to reduce their personal carbon dioxide emissions, we would cut 1 or 2 percent of such emissions. In other words, almost nothing. If you really want to make change, do it this way, he says: put some solar panels on your roof, buy an electric vehicle, then go vegetarian. After that, you can ride your bike more and buy your food closer to home. And definitely, always, work toward wholesale systems change. Avoiding plastic straws? He hasn't ranked that one yet in terms of its ability to halt the rising temperature of the earth. But he guesses, sarcastically, it's about 12,036 on the list.

It is astounding, then, to know what the number one enemy actually is. When the Drawdown Plan—"the most comprehensive plan ever proposed to reverse global warming"—was released in 2017, it named unregulated refrigerants as the top target of the climate fight. Try boycotting refrigerators.

The plan was compiled by an international collaborative of researchers, scientists, and professionals. Its objective was to identify the combination of steps that policy makers could take to halt the accumulation of greenhouse gases in the atmosphere and reduce carbon dioxide levels over time. Its authors looked at any and every possibility, examining things like transportation, building, planning methods, energy sources, and livestock grazing. Even seemingly innocuous things—like making it easier for people to bike or walk—got play. The Drawdown Plan lists wind and solar development, the maintenance of tropical forests, and a shift to plant-based diets among the top ten solutions to the climate crisis—in other words, the very places Nick suggests individual action can really make a difference. But many of the proposed solutions target problems that lots of people don't even realize are problems. For example, regulating food waste is among the top ten priorities identified, with scientists having found that decomposing food is so abundant it is responsible for 8 percent of global greenhouse gas emissions worldwide.

Nick gets most of his information from a diversity of other sources. The gist of many proposals, he says, "is all really about what's economically practical to do first. You do the economically practical stuff and then you do the increasingly impractical stuff." If he were familiar with the Drawdown Plan, likely he would take to it like a monkey in a banana tree. It is full of math.

Drawdown offers a kind of tablature with its proposals, numbers listing possible carbon reductions in gigatons, along with the costs of deploying the ideas in US dollars, then a financial prediction of what doing nothing might cost. The complicated math of it is enough to upset my tiny brain. So fortunately there is a narrative, the one that names unregulated refrigerants as public enemy number one.

Reading this, I felt a kind of shock.

Who could have guessed the sneaking subversiveness of the refrigerator? I had known it only as a place to put food and to avoid getting trapped in, thanks to the childhood PSAs that warned of refrigerators roguishly disposed of in fields. But disorganized disposal was meanwhile also ushering in the end to us all. Turns out, since the world ended the use of ozone-eating chlorofluorocarbons and hydrochlorofluorocarbons (remember when we hated them?), the hydrofluorocarbons that replaced them traded ozone degradation for big-time atmospheric warming. Before 2016, there wasn't a way to dispose of this new enemy that didn't warm the atmosphere a few thousand times faster than carbon dioxide. They were used, as well, in air conditioners, which was ironic since the developing world was already responding to the climate crisis by buying more air conditioners.

Home Depot was, in fact, betting big on air conditioners. Executives said as much in response to a survey by the international nonprofit CDP about the company's climate readiness. This response, among others, somehow qualified Home Depot for billing as corporately responsible, or sustainable, or whatever it was we were calling environmental profiteering these days. In another six months, this fact would inspire the twelve-year-old fellow Coloradan who would cofound US Youth Climate Strike to spend one of her early demonstrations picketing not the state capitol but a solitary Home Depot in Denver. Because fuck Home Depot. But chiefly, it is the refrigerants that menace us.

None of the ideas proposed by the Drawdown Plan are as easily adopted as a straw boycott. And most need government support and a lot of money to implement. But the good news is that it's not really our fault. Which is what everyone wants to think and what Nick wants you to think too—to a point. The point at which you no longer feel overwhelmed and just want to care enough to make sure that your representatives care enough to prioritize fixing things along with all the other "big, capital-R Responsibilities" that the people in charge and you yourself are tasked with. This does not mean giving up electricity, turning the clock back to the days of churning one's own butter, or shearing sheep out in the yard. Solutions are not about self-sacrifice, even though American life must change. And Nick is keenly, acutely aware of how much he needs to stress this point, and of how easy it is to alienate people when he does not. Self-preservation and protectionism can be reflexive in this conversation. And it's important that people understand that when he says he and his co-plaintiffs are doing this work, he is not saying that others ought to be doing it, too, or that it is their fault, or that they have to give up modern life to support the youth.

"That's not what we're saying." The straws, the coolants, the electricity? "All we are asking for is a change in the ways those are made." They're not asking for people to quit things. Or for people to accept the reality of climate change just so they can feel guilty as they go about the daily business of being a person. "If you're going to do anything, you have to think of changes of global effect. . . . The only way you can really change the system, the only way to do it, is by asking for systemic change."

This does not mean that consumer choice doesn't matter. People have influence. Voters have influence. And "if every single person in America said, 'I want clean renewable energy and the only thing I am going to do to accomplish this is to vote in people who will make it happen,' the entire climate crisis would probably be solved in like a year." But that people otherwise find this problem so big that they only either feel overwhelmed or take action too trite to matter is almost as large a problem as climate change itself.

CHAPTER FOUR

"You Don't Have the Right to Tell an Oil Company No"

Whatever is in the way of change, it isn't just about individual denial. American paralysis on climate change is something that trickles down from the very top of the nation's leadership. To an extent, I already know that. But the reasons why don't crystalize until I climb into the car with Jayden Foytlin—the plaintiff whose home has flooded twice thanks to climate change in what were both, ironically, dubbed thousand-year floods—and her mother, Cherri. From Rayne, Louisiana, about 18 miles west of Lafayette, we drive down the highway east into the swamp, following a pickup towing an aluminum skiff.

It's three weeks before trial and Jayden's in the back seat in jeans and a blue windbreaker, quiet with her gaze set out the window. We stop for gas and bathrooms at a place with jojos and fried chicken under heat lamps. Cherri visits a nearby bakery—they make the best apple fritters in the world, she says—and we drive south along a long, straight gravel road somewhere in the Atchafalaya Basin, the largest wetland swamp in America. There's water on both sides and a mound of land piled up between lanes. Cherri drives swiftly over the gravel as the landscape gets increasingly boggy. Her arm rests on the steering wheel, sporting a tattoo: L'eau Est La Vie. *Water is life.*

We arrive at a boat launch, a spot of dirt with a ramp at the water's edge, and with the people in the pickup—a pilot and two activists—we drop the skiff quickly, then get inside and go. It's an overcast morning, warm and thick with clouds, but the water is kind, flat, calm. All around it, the bayou is nature like I have never seen. Cypress dripping moss in the cool muddy water, the riverside washed in rich green. A house here and there, the clapboard weathered and sliding. Logs floating on a soft current. Jayden's hair blows while the boat picks up speed. She has a hibiscus flower tucked into her muddy boot, a gift from her mom, its white petals softly ringing burgundy, white filaments. At the bow, the two activists use arm signals and shout to direct the pilot around floating branches. It's beautiful and—except for the noise of the motor—quiet, until we come around a bend and descend into a pile of tugboats and a barge full of pipes. I can hardly believe it. Orange-clad workers in hard hats are everywhere.

The pipes on the barge are wrapped in giant rounds of weighted concrete, waiting to be sunk into the swamp. This is construction for the Bayou Bridge Pipeline—the terminus of the oil-ferrying Dakota Access Pipeline—and it crosses right through this pristine landscape. It isn't just the stark incongruity that strikes me, construction in such an unspoiled place, or the hideous rift it will make in primeval water, but something else: there on the dock is the sheriff's lieutenant, arms folded, providing security for the workers. Cherri has been arrested once for protesting this project. She tells me clashes are typical among activists, workers, and police.

This realization is sudden for an outsider like me. That in this place, to dissent is to fight. The families here wonder if their men will come home at night, and there are stories of how when things go wrong on the oil rigs and people jump off, they cross their legs and hope for the best, despite the pods that are supposed to keep them safe in such emergencies. If you do it wrong you're dead, they say about this leaping. But then again, if you're doing it you're probably dead anyway. Still, oil is the law of this land. It is its own order. The sheriff is just the sheriff. And this is what the *Juliana* plaintiffs seek to dismantle—an order as old as the Rockefellers, one that stretches from the stock exchange to the Louisiana swamp.

Jayden is next to me in the boat while all this sinks in. We've hightailed it around, and her long, dark hair is blowing on the wind again as the rain starts to fall. Her expression is a mix of emotion and resignation.

"What do you think when you see that?" I ask her.

Her words come slow, measured. She says, "I'm very upset and kind of frustrated about it. It's really gross and disgusting just to see it, and especially to see the workers just working on it as if it's nothing."

I ask her if it's just this pipeline that bothers her or if it makes her think of other things.

"This isn't the only pipeline," she says. "The same company is doing it in other places. The Dakota Access Pipeline? Same company. They know they're putting it in very sensitive parts of Louisiana. Our rivers and bayous are at risk, and they know that. They're doing that but they're like, 'Well, you gotta make money.' But most of these aren't even permanent jobs." The workers, she says, mostly come from Texas.

She's right. The pipeline is 162 miles long, connecting oil refineries in Louisiana to an oil and gas hub owned by Sunoco Logistics and Phillips 66 in Texas. In between, it bisects the Atchafalaya Basin. Its cargo—crude oil—is intended to complement the two pipelines carrying Bakken crude to the Gulf Coast from the north: the Trunkline Pipeline and the Dakota Access Pipeline, famed for spurring the Standing Rock protests in which other Native American resisters were attacked with water cannons and dogs.

I'm surprised that Jayden knows all of this. She's fifteen, after all. But then again, there's a Standing Rock resister spotting on the hull of this boat. And I'm consistently surprised by how much the plaintiffs do know about the fossil fuel industry. When I ask them what strikes them most about their co-plaintiffs, almost everyone who isn't Jayden talks about Jayden. About how hard it must be to be a plaintiff in this lawsuit and live in Louisiana. To be in high school, have a dad who once worked the oil fields, and be the kid that doesn't go in for this culture—the strength it must take.

Despite the impact of climate change on her community—torrential rain and rising waters have flooded the region and other parts of Louisiana—and that pipeline construction degrades the wetlands designed to absorb the water, Jayden has been derided for her role in the *Juliana* case. When her best friend's parents found out about the lawsuit, Jayden says, she lost her best friend. Kids heckled her at school. One joked that she didn't have to do her homework because she was suing the federal government. People gossiped about her in the cafeteria, made fun of her, called her brainwashed and part of the liberal agenda. Some asked her what was wrong with her. One boy made comments about her mother. Jayden thought that was off-limits—talking about somebody's mother.

"I didn't want people to know," Jayden says, adding that kids learned she was a *Juliana* plaintiff from an interview. "I was right when I said I didn't want people to know, because it ended up this whole big thing of people—like, the conservative students or people that had that mindset—would challenge me in the middle of classes with political debates. And I'm just trying to do my science work. I don't need all of this."

In this way, Jayden's story is a microcosm of what's stifling change in America. Like the story of the *Juliana* case so far, it makes clear how deeply the fossil fuel industry is entrenched in our communities and economies, and thus our government too.

In the morning, I'd first encountered Jayden in her bedroom—purple—after I found my way down a dark hall of her pink house. Jayden's older sister Erin, a few minutes awake, was flopped on the bed under a pastel comforter, and Jayden was at her desk, the wall above it covered in watercolors and drawings. Jayden was friendly and talked about her activism and her art and her love for aeronautics. Told me how captivated she is by planets and how she hopes to one day study science. She said the planets that used to adorn her room were still put away. The reference was to the flooding that figured large in her pleadings. So I asked Jayden to tell me about the floods, especially the flood in 2016, and her expression darkened.

Here's what she said: "I woke up at like three, somewhere in the a.m., because ... my older sisters Erin and Grace were knocking on my door and telling me to 'Wake up! Wake up!' and that water was coming through the house. Of course, I thought I was dreaming because I was like, Why the hell would water be coming through? ... I was kind of thinking ... What do they want? I thought they were just trying to make me come out of my bed with some excuse or something."

They weren't. Jayden sat up. She swung her legs to the floor and stepped into water up to her ankles. The house was a ranch on a slab, an L-shaped building with the longest length reaching deep into the backyard. Jayden was at the end of this stretch, in a room with a door to the yard, and outside it was raining hard. She walked toward her sisters' voices in the hallway, opened the door to find them there. And as she did this—pulled the door toward her—water rushed from her room into the rest of her home.

"I was like, Oh that's my fault . . . oh my God, why did I open the door?" she said.

Elsewhere water had been rising through the cracks in the foundation, soaking the carpeting. Sewage and old water rose in the toilets and in the bathtubs. But until then, most of it was in Jayden's bedroom with Jayden, slowly leaking in from under the door.

The sewage "had the worst smell in the world. It was so nasty," she said. "My family ended up getting sick. A lot of my neighbors ended up getting sick."

Jayden's brother, Dylan, her sister Grace, and Grace's boyfriend went to the police station, but no one could help much, just gave them a few sandbags to block the doors while the water rose. It wasn't enough. So next they used all the blankets. Then the towels. So much water, they were no match for it. They gave up and focused on the electronics, on unplugging everything, worried about electrocution, and trying to save what could be saved. Drawings and photos were lost anyway. Jayden's bedframe dissolved in the lake of her room. Her brother's toys were next. When I asked how high the water rose, she pointed to a watermark on the wall about 2 feet above the floor.

"I hate that flood so much," she said.

Sometimes when Jayden talks, her arms and hands move in a way that could pass for dance but really is just pent-up energy that leaves her through her hands. She did this as she told this story, moved her hands. Trauma looking for a way out.

Erin threw up for two months. Both girls are Diné and Latinx, with some German and Cajun mixed in, and Jayden said Erin went whiter than her skin ever was. When they took her to the doctor, the doctor seemed overwhelmed. So did the FEMA workers who arrived to inspect things later and to cut a hole in the wall to check for mold. There was a lot of it, and it cost Jayden's family a bathroom. They threw out their furniture, the carpet too. And when the subfloor started to peel, its tentacles threatening their feet, they stripped that also, set up house again on top of the bare concrete. Two years later, they are still living on the slab, slowly reclaiming the house with the piecemeal paint and fixes they can afford after FEMA money only paid to remove the hazards. They didn't realize they didn't have flood insurance until they needed it; floods used to be unlikely where they live.

Ever since, Jayden said, she worries when it rains. "I don't want my room to be ruined again because it's still not even fixed," she said. She still needs a bedframe.

The torrential downpour that flooded her home turned out to be the nation's worst natural disaster since Sandy, dumping more than 7 trillion gallons of water—three times the amount of rain that fell during Katrina—and flooding about sixty thousand properties in all. Scientists quickly tethered the unusual rains to climate change, finding that rising temperatures likely increased the storm's rainfall by 20 percent. They also warned that the Gulf Coast conceivably faces a future of "nonhurricane-related, warm season extreme precipitation."

This kind of rain plus coastal flooding has been ongoing since. But while we drive home from the swamp, Cherri tells me that despite the flooding, the people here don't take issue with climate change. They take issue with the Army Corps of Engineers—the government entity that's supposed to manage the water—and with local officials. And indeed, in online forums, there are calls to dredge rivers, shore up levees, curb development. Jayden was already a plaintiff in the lawsuit when the flood hit; she joined after meeting one of the attorneys at an action for hurricane relief for Florida. Now the floods—this one and another a year later—strengthen her case.

"Some people said, 'Yo, that's really cool.' But there's a lot of people that were like, 'What is wrong with you?' 'You're being brainwashed,' and stuff like that," Jayden says. "I don't even identify as liberal."

Later, Jayden's mother, Cherri, a longtime activist and the director of Louisiana Rising, says the way Jayden has been treated is indicative of how righteous people feel in the farthest reaches of polarized politics. And how justified they feel, too, in directing their aggression at anyone who doesn't share their views.

"She's just a little girl. People feel so okay ... with like going after and talking that way about a little girl," says Cherri. "That's how safe they feel in their opposition and in being ugly."

It's not surprising that Jayden's community is frightened by her lawsuit. What a *Juliana* victory would entail—a restructuring of the way oil and gas development is authorized, permitted, and subsidized in the United States—threatens the bedrock of the Louisiana economy.

This culture, the one in which oil is Louisiana and Louisiana is oil, runs nearly as deep as the history of humans and oil itself. Like a lot of stories about the industry, it involves a drill, a gusher that rained oil, and a field of ruined crops. It dates back to 1901, when an intrepid twenty-something named W. Scott Heywood dared drill a well deeper than any other, having calmed a nervous rice farmer named Jules Clement enough to secure access to his land. Ninety miles away near Beaumont, Texas, and nine months earlier, Heywood had drilled wells at Spindletop, the legendary geyser that seeded commercial oil in Texas. The Louisiana well he drilled that soon struck oil, thereafter known as Clement No. 1, was unlike Spindletop in one important way: it was drilled in what industry lingo calls an anticlinal trap, a spot where non-porous rock lay over where oil and gas and water stewed. Which meant the Clement No. 1 well tapped a reservoir of oil, not a small pool, realizing the vast reserves of fossil fuels undergirding much of Louisiana.

The Clement No. 1 well became just one well in the Jennings oil field soon after. Located in Acadia Parish, where Jayden lives, it was a harbinger of things to come. Within a decade, a federal report on the oil bounty in Louisiana numbered 215 pages, stuffed with accounts from every parish. Jennings, by then, was producing so much oil from so many wells, a map of the site looked like a dense dot matrix, if a bit off alignment.

The oil field peaked in 1906 at 9 million barrels, enough energy to power the average American home for 550,000 years or fly from London to Louisiana more than 1.7 million times. The surrounding land, described then by federal officials as "monotonously flat . . . a portion of the Gulf floor that has been raised . . . and is now slightly dissected by sluggish meandering streams," was basically pay dirt. That it all used to be underwater, part of a wide swath that stretched north from today's Gulf Coast into a triangle to Illinois, made Louisiana's fortune. Beneath it were the classic geologic features of oil beds: Quaternary and Miocene sands that inevitably fossilized a sea of algae and zooplankton.

Through this luck of natural resources, Louisiana is today the number two producer of crude oil and the number four producer of natural gas in the United States. It ranks second in the nation in petroleum refining capacity, with eighteen oil refineries. The state itself and the waters off its shore are crisscrossed by more than 92,000 miles of pipeline. And this trifecta of wells, refineries, and pipelines supported $72.8 billion in sales in 2015, along with $19.2 billion in household income and 262,420 jobs. In Acadia Parish, the energy industry paid nearly 12 percent of the property taxes collected there and more than $22 million in local wages the year before the flood that inundated Jayden's room.

This relationship between the economy and the oil is older than anyone alive in Acadia Parish can even remember. So, too, the relationship between the oil and its investors. When the Clement No. 1 well struck oil in 1901, investors who'd bought in at $1 a share and held on made huge profits, having gambled better than those who lost confidence and sold for 25 cents a share along the way. Today, oil and gas are still the primary energy sources powering the world, and thus, still largely powering the US stock market. If you look at the total market value of the companies included in the S&P 500, 11 percent of that value comes from the energy sector.

But it isn't just the stock market that matters in explaining why fossil fuels have such a stranglehold on America today. It's that somewhere along the way, after the United States fueled its allies through two world wars, built the most technologically sophisticated military on the globe, and powered its own massive economic growth on cheap oil, it developed a habit it couldn't break. Postwar prosperity brought cars and roads and plastic to a world economy in which Americans rode the crest of a wave. The only thing our nation lacked was enough oil to keep it all going. Which explains why the United States swaggered its way into midcentury deals with Saudi oilmen and assumed the inevitable peacekeeping (or was it oil-securing?) missions in the Middle East that came later.

Once America's dependence on Mideast oil took hold and a certain global order followed, the United States became a kind of de facto global police force in exchange for its access to oil. No one explains the delicate geopolitical dance it takes to maintain this order better than Paul Roberts in his 2004 book, *The End of Oil,* which remains a must-read for anyone interested in understanding how a pampered Saudi prince could be responsible for the killing of a US-based journalist and walk it off with barely a public dressing-down in 2018. It is through these geopolitical arrangements, and a certain neoconservative ideology fostered by two Bush presidencies, that an attachment to oil set deep roots in our nation's culture. But more critically, it is through these arrangements that America's place at the top of the world order is held secure. To believe in a world without that order is not only to relinquish power but also to believe that peace is possible when we are not in charge and when we are not the richest nation in the world. It's a tough sell at a time when world resources are only getting scarcer and when conservative values trend toward self-protectionism. Viewed through the lens of the breakdown of world order, climate change is the more distant threat.

This helps explain why in Louisiana, maintaining that order is synonymous with maintaining the oil and gas industry. And how difficult it can be for someone like Jayden to stick her neck out, and what she and other people who oppose fossil fuel development are really up against.

When Energy Transfer Partners, the pipeline company that built the Dakota Access Pipeline at Standing Rock, came to the state to build the Bayou Bridge Pipeline, regulators refused to permit its usual security detail, a company called TigerSwan. The decision was supposed to promote the peace. It was TigerSwan that was responsible for using attack dogs and spraying protestors with water cannons at Standing Rock, tipping the resistance effort into a full-tilt civil rights standoff, one that galvanized people all over the world. But instead of hiring another security company, Energy Transfer Partners let a passel of security contracts to local law enforcement, setting off a civil rights issue of another kind when officers blurred the lines between the job of protecting the public and the job of protecting the pipeline. State patrol and probation officers, local law enforcement in St. Martin Parish, the place where the pipeline crossed the swamp, showed up to these off-duty second jobs in pipeline security in marked patrol cars wearing police uniforms and carrying service weapons.

Thirteen protestors were ultimately arrested. They included Jayden's mother, Cherri, who by then had founded L'eau Est La Vie, a camp for resisters, and was leading activities as apparently subversive as performances of *Crawfish: The Musical* at construction sites. Those arrested said they often weren't clear why they were arrested or by what agency. Some say they were dragged from the water—which is a public right-of-way—onto construction sites, then charged with trespassing. Some landowners had given permission for resisters to be on their land. Ironically, several had never sold their land to Energy Transfer Partners for the pipeline, but construction crews marched forth anyway, digging trenches and burying pipes.

After arrests, at least one police department posted photos of the protestors on Facebook and other social media, making them ready targets for single-minded trolls who threatened their lives, their loved ones, and their homes, among other things. When they targeted Cherri, she received threats that her daughters would be raped and that friends staying in her home would be killed in her front yard.

This intolerance and speech squashing is true all over America. Twenty-one states have adopted or considered legislation to worsen penalties for those who protest "critical"—as in "fossil fuel"—infrastructure. Only five states have rejected such proposals. The Trump administration has proffered similar federal legislation that makes protesting, or "inhibiting" or "disrupting," infrastructure, or attempting to inhibit or disrupt infrastructure, a felony punishable by twenty years in prison. Louisiana was among the states to adopt such legislation.

Elly Page, a legal advisor at the International Center for Not-for-Profit Law, says the legislation seems intended to have a chilling effect on pipeline protests. Many of the bills significantly heighten penalties for conduct such as trespass onto pipeline property or interference with pipeline construction. The penalties are in some cases draconian, she says—under North Dakota's "critical infrastructure" law, for instance, interfering with pipeline construction is a Class C felony, punishable by five years in prison and a $10,000 fine.

Hypothetically this could capture, for instance, a protest that blocked an access road to a construction site, Page notes. And almost all the proposed and enacted laws penalize not only individual conduct but also the people and organizations that support that conduct. "In effect, these laws dramatically raise the stakes for individuals involved in protest activity," Page says. "People who want to exercise their First Amendment rights now have to fear being caught up in these overbroad and vague laws with extreme penalties, and they may decide that it's not worth the risk."

Cherri summarizes the status such legislation gives to corporations: "It's not a felony for me to go to the White House and to protest there. So what they're doing is they're even openly, by law, even putting these oil companies above the entire United States of America in the amount of protections that they have," she says. Criminalizing the opposition makes it easier to throw its members—and its message—in jail, shut them up. "People will only let that go down if there's a criminal element there and if the police are able to come out and say, 'Oh yes, those are criminals.' Then nine times out of ten, the public, in their ignorance and tomfoolery, will just sit back and say, 'Oh, okay. Well then they probably deserve that.'"

In summary, she says, "You don't have the right to tell an oil company no in 2018."

On the ride back from the swamp, Jayden sacks out. She puts her headphones on and stares out the window until she is lying down and if not sleeping, close. It was an early morning, and owing to the day's jelly beans, a sugar crash is probably imminent. Jayden strikes me as being this way: two speeds. When she is into a thing, she is animated and talks fast. But when she's not, she's quiet. Watching and thoughtful. She listens to music, probably K-pop, and watches the live oaks and the dry lawns go by, maybe one ear open while Cherri and I talk up front.

Two days earlier I'd driven these roads outside Rayne in search of where the land meets the sea, south of Jayden's hometown and into the White Lake Wetlands, trying to understand the rising waters by finding the edge of the land. That's what I'm used to, after all: an edge of the land. A place that ends in beach or rock or sometimes both.

A habitat between civilization and water. A boardwalk, a cityscape. But that doesn't exist here.

Instead I drove through land so flat it seemed to bend the curve of the horizon upward. Past the signs for po'boys and zydeco dance and the tractors for sale and the sugarcane fields. Past the billboard with the towering cutout of a lawyer in a suit, promising payouts for oil rig accidents. I drove until the dragonflies got so big I flinched when they neared the windshield.

Along the way, I learned things. Like how the storm clouds gather at noon and sometimes issue only hollow threats but other times break open so a massive amount of water falls out. And how afterward there are a few minutes of reprieve, but otherwise the air is unceasingly sticky hot. And how people jaywalk as if there's no time to spare before melting. And drivers speed madly down seven lanes, sequestered in their air-conditioning until they can park and dash to the other air-conditioning. Grocery store, pharmacy. Many of these places smell submerged, as if they've been dredged out of a fish tank.

I drove until the horizon further inverted itself and fuzzy little trees dotted the line between land and sky. I drove until *the Sultans play Creole* and the buildings started to rise up on stilts. To where irrigation ditches were full and there was moss growing inside and dikes were required and cattle were nosing around in between. Past a cemetery where people get buried above the ground, and flagpoles, too, along the way. But I never reached the end. Never a rocky outcrop or that patch of sand. Just a steady negotiation between land and water until the water started to lap at the sides of the road.

East of the wetlands, there were turnouts now and then, short dirt roads on levees between water and farms, the occasional boat launch into a canal. One after another, these roads were full of garbage. Gatorade bottles and plastic bags. Glass and cans rolling on the ground. No one to come and clean them, it seemed. Which speaks to how these wilds are held. Farm gate, cow, trash pile, crop. Utility poles riding off into the distance. All the while the radio told a story about a haggard man come into church, the devil inside him, until Jesus commanded and the demon threw the man away. It all went on long enough that the landscape started to blur into American flags, mythic evil, and garbage. Soon the birds and the bugs came to take it all. Ibis and cormorants, vireos and wrens, the ever-larger dragonflies riding on the wind.

Eventually I reached a place called Pecan Island, found a trailer on pilings, then a house on 10 feet of stilt. There was a gas facility near where it all just sank into the sea. Kinetica's barge terminal for liquefied natural gas. A few miles west, a preserve named for the Rockefellers.

Cherri asks me about the case. If I know what will change now that Kavanaugh is about to be appointed to the Supreme Court. If I think a conservative court will affect *Juliana* and its trial date in three weeks. This is a question a lot of people are asking lately, but I don't know the answer to it. No one really does. Kavanaugh has just undergone the first days of his confirmation hearings, and Cherri says the politics of his confirmation make her mad, that it is especially intolerable to watch mothers of daughters reflexively defend him just because he was picked by a red president. Now another thing to protect with all the rest: her daughters' sense of being girls.

We pass a campaign sign for Brian Douglas Theriot, who is running for sheriff. Cherri says there are only two things you need three names for: being a serial killer and running for office in south Louisiana.

The route back takes us to the L'eau Est La Vie camp, where it is immediately clear the effect that all this downward pressure from industry has on the people resisting it. And how Jayden finds a home as a lone teen oppositionist.

On this patch of land in the Cajun prairie, only a row of tents, two metal-roofed structures, and an inflatable swimming pool seem to fend off the heat. Cherri pulls the car into a muddy parking lot past a steel gate. The pipeline was supposed to go through this land, but it won't now. Rather than fight the camp, Energy Transfer Partners went around. By the gate, I see a handmade catapult on a wooden frame, built from rubber cords and what looks like a hefty colander, used for launching projectiles amid hostilities with construction workers.

People are cautious when I turn up, trailing Cherri and Jayden along a muddy path to the larger metal-roofed hangar. Roughly a dozen people are around, one making a meal at a camp kitchen, some painting banners, and others sitting in a circle of chairs, the conversation stalled in the chill I seem to have brought. I don't get the stink eye, but there are sideways looks and a lot of quiet. I can read the collective thought—*reporter in the house*—and it is clear how nervous I am making them. High up on a wall is a list of rules that define the social order, a list that also makes obvious how much the usual civic order is not welcome here. No calling the police, for example. No unauthorized guests or photos. No saying who else is here. There are details of a tribunal for hearing crime reports, instructions for a security system that amounts to a patrol, and rules for interacting with the press.

Jayden finds a chair and settles in. This is our second visit of the day, and earlier she'd grabbed a plate of eggs and a cup of coffee with flavored milk and whipped cream. Now, she chats easily with the folks in the camp, so it's clear that she fits in, knows some of the people here. But most of the conversation is adult conversation to which Jayden is privy and listens quietly, Jayden-style. She crosses her legs and leans back, denim and T-shirt in the sticky sun, seemingly indifferent to the heat.

The group is a collection of shaved heads and curly hair, funny T-shirts and phone fiddling, cutoffs and busy dresses, fitness and wild beards. Only two decide they are comfortable enough to tell me their names. Some go by alt names, like the guy with the funny T-shirt who gives me the name of an animal in place of his own. Their conversation tells me they are hobbled for cash, only rarely find a bed, and that separate from their ongoing efforts to resist fossil fuels, they are locked in a second battle with the mosquitos that overtake them in the night.

The room around us is divided into quadrants, and we are in one of them in a circle composed of mostly camping chairs. The next is a kitchen sporting two huge burners on propane tanks, a hotplate, a coffee pot, a mini fridge, plus a dish station with three tubs. The third is just stuff. The fourth includes tents where people sleep in mosquito nets and a speaker that's been playing someone's playlist, first Amy Winehouse, then Paula Fuga.

A guy with curly hair has a question: Does anyone have PayPal or Venmo? He wants to know whether anyone has a way to accept electronic money, if they feel comfortable putting themselves on the grid in that way. He says a guy he knows wants to rent him a hotel room, but he doesn't want to deal with the electronic side of that. There are suggestions and debate. Then Curly Hair says his girlfriend is upset about the vagaries of his schedule. He says he has a three-day window during which he can account for himself. And that he doesn't want to have to plan farther ahead than that. He might need to move, to beat feet from this camp or another, so it doesn't make sense for him to commit to more. A woman says she would have a problem with that if she were the girlfriend. She wants to know whether this is about the scheduling or the commitment. Curly Hair is momentarily stumped.

Outside the door—one that rolls up like a garage door—is a rain barrel with a tray of seedlings on it. Behind that is a wooden house tucked into the prairie like a photograph I will never take—no photos allowed.

A small group is painting a banner. I pick up on the details of an upcoming action but no one wants to provide me with the broad strokes. Above it all are banners with frogs and turtles that work as a kind of radical decor. When I ask what the camp is for, what the campers do there, people shake their heads or refer me to Cherri.

In the lawsuit's early days, it was mostly Big Oil that stood in the way of *Juliana v. United States*, stood in the way of an American plan to reduce global warming, much like it stands in the way of these resisters now. When the case was filed in 2015 and Obama held the White House, he had already given the State of the Union Address in which he called climate change "the greatest threat to future generations" and signaled his intent to sign the Paris Accord.

By then, Obama had spent $92 billion in stimulus money seeding an alternative fuel industry in America. So the fossil fuel industry did not test whether the Obama administration would defend the nation against the *Juliana* plaintiffs. Obama was a president who had opened vast public lands to windmills and solar panels, set goals for renewable energy development and alternative fuel standards for cars. He'd also protected huge swaths of ocean in part for carbon sequestration, was slashing carbon emissions from power plants, and in doing so, squeezing coal hard. Though Obama also supported fossil fuels, spurring massive upticks in domestic oil and gas production, industry leaders did not wait to see whether Obama's Department of Justice would defend it against the *Juliana* twenty-one.

Under Obama, the Department of Justice did defend against *Juliana,* and with many of the same legal roadblocks later deployed by the Trump administration—filing motions to dismiss it on the same constitutional grounds that the Trump administration would also pursue. But with Obama at the helm of the nation, it was the fossil fuel industry that fought the *Juliana* case hardest, taking aim like a bull at a toreador. Beginning in 2015, fossil fuel corporations sought to become defendants in the case and did. Then they filed brief after brief challenging the merits of the litigation. For the next two years plus, industry attorneys waged a furious paper war. It was straight out of the Big Oil playbook. Where they could not outlitigate, they could outspend, harass, wear down. The American Fuel and Petrochemical Manufacturers led the charge with the American Petroleum Institute, representing the biggest of them: Citgo, Chevron, Phillips, and ExxonMobil, to name a few, the latter having spent heavily—along with the Koch brothers—on seeding climate-change denial, not to mention having ties to nine out of ten of the most prolific authors casting doubt on climate change.

The list of fossil fuel companies that stepped out to battle *Juliana* reads like an S&P industry index. Led by their trade groups, they sought pretrial appeals, attempting to get the case dismissed, and filed objections to the production of documents and to shield former Exxon CEO Rex Tillerson, who at the time was serving as secretary of state, from being deposed. Then, when Trump assumed office, they stopped. There was a question about whether the companies would have to produce documents if they stayed in the case. And Trump's cabinet represented them well enough without the added expense of their own attorneys. Trump soon proved as much by nominating a former BP attorney to head his administration's defense of environmental matters, including the *Juliana* case. With Scott Pruitt, Ryan Zinke, and Mike Pompeo at the helm of environmental policy, and their deputies, too, the fossil fuel industry was enjoying a nepotistic heyday it had not had since the golden Bush years. Many of these gains included supplanting departmental deputies with industry lobbyists. Seemingly overnight, they slipped under the door of the EPA and the Department of the Interior too. David Bernhardt. Frank Fannon. Mike Catanzaro. Sean Cunningham. Too many to list. All were disciples of the same corporate complex that had vigorously battled *Juliana* when the case was first brought.

It was fitting in a way, then, that Trump's ascension to the White House coincided with a universal withdrawal from the *Juliana* case on the part of the fossil fuel industry. ExxonMobil was, after all, among the last vestiges of the old Standard Oil Company. In other words, the last vestiges of Rockefeller money. Who better to defend a Rockefeller than a Trump? Even in the days when the Rockefellers had started embracing the climate cause, it was a moment among many to illustrate that the rich were still in charge. And to demonstrate who many of them were: oil barons and natural gas tycoons.

I take a walk to a porta-potty and leave the boots I borrowed in the other hangar on the way. It's a big enough space for a skiff to fill half of it. On the other side is an inflatable pool, empty today, and behind it a clothesline full of odds and ends of fabric and clothes.

I don't know what happens in this camp. I don't know whether any of the direct action against the oil and gas industry is illegal. But after I leave, I never hear any news from Louisiana that makes me think so. What I see instead are mostly innocent things like banners, and the strongest claim that can be made about any of it later is the claim of trespassing and work slowdowns because of things like *Crawfish: The Musical*. Which makes it that much clearer that in this environment, where one's biggest subversion is to dress like a crustacean and dance in front of a bunch of bulldozers, even an act as benign as a theatrical protest—one that would fly easily and regularly in a place like Eugene, Oregon—feels dangerous enough that the performers take care not to speak one another's names, and hide out for the time it takes to prepare for such a thing.

For all the opposition to the opposition, by the time this pipeline is complete, there will never have been destruction of property here, nothing more unsavory than a few projectiles between camps and some extended Facebook Live events starring people costumed as swamp creatures, their faces masked with bandanas. For that little bit of speech to inspire this kind of cloistering, of fear, is a fright all by itself.

An eager puppy greets me, and somehow Jayden and I end up standing in the mud of the parking lot, talking about boys and friends. Her new best friend is Miko, another plaintiff in the case, and in this moment like the rest of the day when she's talked about her co-plaintiffs, she runs through reasons why each is special and unique. Many are good speakers, she says, which she envies because she often lapses into slang and has to focus on not swearing. With these plaintiffs, though, Jayden, like Alex, describes a comfort and a peace she doesn't feel with many other people. She tells me about a camping retreat in Oregon some months back, a deeply bonding experience to prep the plaintiffs for trial. They saw a meteor shower from the Oregon wilderness, something she can't see from the light-polluted Rayne sky.

"It was so cool," Jayden says. "You can see the ring of the galaxy above us and there's so many stars." She is looking forward to spending time in Oregon for the trial in three weeks, for the trial itself and for the opportunity to see such a sky. She's not sure what kind of science she wants to study at college, but she is thinking about astrophysics. And art too.

When she met the other *Juliana* plaintiffs for the first time, Jayden says, "I was a really, really shy kid.... I was so scared. I was just extremely scared. But now I'm way more confident in myself. Which is really good. I'm not just more confident in myself but also in the work I did. At the time I thought, 'What if the other plaintiffs don't like me?'" All that self-doubt, that has changed.

Jayden believes in the lawsuit, but there is more in it for her. She finds a sense of community in it, and in knowing there are other young people like her, people who want the world to change. To see herself move from that shy girl who couldn't talk, the one who used to be heckled in school, to the one who knows she belongs, who handled her deposition like a pro, who is proud of herself and of her activism and lives with purpose and a sense of belonging—it is its own return, immeasurable.

Lately it feels like power to stand up. Against Big Oil and against that flood that she hates so much. For her damaged room, her gone bed, her brother's toys, and her sister made sick by it all. For the camp. For the people who stand with her.

CHAPTER FIVE

Radical Corrective Streak

Oregon is, of course, a great place for a resistance. By the eve of the trial, the state had the quirky reputation of having been chosen as an oppositional stronghold by a whole bunch of other renegades. It had been the home of Rajneeshpuram (the community of orange-clad religious devotees who made their own city in high desert until they poisoned a few salad bars) and the frontline of the takeover of Malheur National Wildlife Refuge by antigovernment types bent on free cattle grazing and other militant notions of land management. But these rebels and standoffs weren't what made Oregon an ideal staging ground for *Juliana v. United States.*

The reason for that is Oregon's legal history. Oregon's district court is in the Ninth Circuit of the federal court system. And by October 2018, the courts in the Ninth Circuit already had experience doing what the *Juliana* plaintiffs wanted them to do. They had intervened in disputes and acted as an intermediary between the wronged and the other branches of government. A key illustration: the Northwest Fish Wars, years of treaty rights cases in which the district courts of Oregon and Washington had helped tribes, states, and federal agencies work together on plans to protect salmon, plans that were supervised by judges. In the words of Mary Christina Wood, the thinker behind the legal theory underpinning *Juliana*, these cases are "the absolute best example of a judicial remedy." They are better even than cases involving segregation or prison crowding, or any other case in which judges have intervened in executive operations.

It didn't hurt, either, that the Ninth Circuit Court of Appeals was comprised of judges who were denizens of the West, and two of the judges were from California, where extreme temperatures were already fostering drought and wildfire that was far more serious than in most other states. The court had already weighed in on California's aggressive efforts to regulate pollution and environmental toxins. And Judge Ann Aiken, the judge in the US District Court of Oregon, "completely understood the type of case it was, the urgency," Wood noted. "She said, 'This is a civil rights case.' So she was talking about institutional remedies."

These kinds of ideas were in sync with the place from which they sprang. Embracing the unconventional and fighting for natural values have been hallmarks of Oregon life for nearly a century. Deep care for the environment is not a partisan thing in this state. Instead, it is a lesson learned from the days of Manifest Destiny, when Oregon's bounty was something that capitalism exploited until it occurred to its citizenry that abundance was a thing to protect.

In terms of its natural endowment, Oregon is a place unlike any other in the nation. Made from ancient tsunamis and plate shifts, followed by millions of years of volcanic eruptions, earthquakes, and floods, the state is like a sampler of the most dramatic landscapes in America. As a land, it is eye candy and splendor, an adventurer's dream from earliest days. Despite the assaults modern life commits against it, the dazzling spectacle of its landscape manages to never be anything less.

Oregon rises out of the frigid, hard-charging Pacific as a series of headlands, rock islands, and reefs. In between lie stretches of estuary, rocky beach, and incomparable dune. Here, mudflats and shore pines quickly give way to riparian forests and the Coast Range, so at points along the 362-mile coast, mountains practically shoot out of the sea, rising so steeply they are visible from miles offshore.

Oregon's eastward border is no less dramatic. It took hundreds of glacial floods to carve the state's gorges and rivers from the east. To the south, those floods left canyons between high deserts, a lowland desert in the dry southeastern corner, a few salt lakes, and a geothermal oasis. Volcanoes, including one called Mazama that erupted into ash and left the splendor of Crater Lake behind seventy-seven hundred years ago, still ride the ridge of the Cascade Range in the peaks and snowcaps that are the backbone of central Oregon and the pride of its skiers, hikers, and backpackers. To the north, a mountain lake keeps the Columbia River and its tributaries flowing, a system once so stocked with salmon that Nez Perce fishers could catch one hundred a day to a man.

Native peoples who lived here for thousands of years found this land had all they needed. They wove baskets and hats from grasses and tree roots, built canoes from towering cedars, and longhouses from frames covered with woven tule and cattail. They traded furs and salmon and buffalo meat and hides, made jewelry from shells and arrowheads from obsidian, and carved tools from pipestone and basalt. White traders and settlers pillaged it quickly, wiping out sea otters and nearly decimating the beaver population in the decades leading up to 1820, all in the name of shiny hats. Logging, mining, fishing, and farming became prime industries early on, and fishers fished so hard that Oregon had its first hatchery to support salmon by 1877. Within twenty years, the federal Carey Act allowed corporations and individuals to divert waterways into farm irrigation. The damming of the Columbia River plunged Celilo Falls—a sacred site for Native fishers and a long-occupied hub of trading and culture—underwater in 1957.

In short, early colonization was such an eventual mess that by the mid-twentieth century the pendulum had swung widely in correction. The federal Mitchell Act paved the way for the Oregon State Sanitary Authority, charged with boosting support for salmon habitats and regulating water and, eventually, air pollution. Fixed fishing traps were banned for salmon, and the state started requiring the timber companies to replant the trees they were cutting. State efforts also dropped millions of seeds over a burn area east of the Coast Range, creating today's Tillamook State Forest. In 1967, it was the state's Republicans, not Democrats, who set aside the entire western shore for public use, the act memorializing Oregon's unique brand of conservative conservationists. The seventies brought the first bottle bill in the nation, also created by Republicans; the eighties, the preservation of the Columbia Gorge.

In between, Mount St. Helens erupted and forests burned, whole towns flooded, and Oregonians learned there's only so much humans can do to exert their dominion over nature after all. Today this radical corrective streak remains in the form of some of the most comprehensive land-use planning in the world, as well as citywide bans on plastic bags, local climate ordinances, attempts at carbon taxes, and state subsidies for renewable energy, electric cars, and wave power.

Of the eleven *Juliana* plaintiffs from Oregon, six are from Eugene, a liberal university town roughly a hundred miles south of Portland and home to one of four divisions of the US District Court of Oregon. With Portland to the north, Eugene anchors the southern end of the Willamette Valley, one of the most fertile valleys in the world, known for its rich soil, its many rivers, and the convergence of the Cascade Range with the valley floor. This particular geological windfall was made possible by massive floods that rose up off the coast of Washington and swept the topsoil off the northern state, pushing it through the lowlands between the Cascade and Coast Ranges and leaving it there. Owing to the dazzling fertility of this soil, farmers in the Willamette Valley now grow 90 percent of the world's organic seed and 22,000 acres of wine grapes, and are notorious for their beer hops and their greenhouse and nursery stock. They also produce most of the grass seed, hazelnuts, and Christmas trees sold in North America. Thus, civic life in the town of Eugene revolves around these things: whole-grain foods, fruits and vegetables and potlucks, beer and wine, things made by hand or strummed, hiking and school. Still, Eugene's wholesomeness is tinged with Oregon's misfit disposition, a place where senior citizens dye their hair purple, a town of people who break things and set trends in a state full of people whose inclination is to go it alone, sometimes together.

Small wonder then that Zealand Bell, a self-described "outdoor person," refers to the meadow behind his home as the place he likes to relax. At age fourteen, the high school freshman, clear-eyed and clad in North Face, can lucidly articulate how nature makes him feel. And how the thing he likes best about the outdoor hobbies that steal his time—basketball and crew, kayaking and mountain biking, surfing, hiking, skiing, and jaunts to the beach—is that they allow him to be close to this feeling. "I just like being out in nature and having fun," he says. "A lot of times when you go to certain places outside, it looks different than the time before because nature keeps growing and changing with seasons.... I just like to take it all in."

In the days before trial, before anyone knows that there will not be a trial, Zealand describes himself as excited to have reached this stage but nervous about the info dump that is about to take place in the courtroom. He has no question about whether he wants to be there. He has been there since age six, when he first learned how climate change affects basic life and the outdoor ecosystems that are dear to him. After a summer camp run by Our Children's Trust, back in the days when the organization helped kids undertake local advocacy, he was one of the few plaintiffs who lobbied Eugene's city council for the most stringent local climate ordinance in the country. When he was asked to join the case later, the yes that followed was easy.

His parents didn't hold him back. He was a child with a strong sense of justice who had not lost interest in what this issue meant to him. And because they are both teachers, his parents also understand what the issue of climate change means to most kids, who learn about it in school and grow up deeply unsettled by the adult inaction that persists as they age. Still, by fall 2018, in the lead-up to trial, Michael Bell and Kim Pash-Bell, like a lot of the plaintiffs' parents, are slowly awakening to the mayhem that trial will soon inflict on their lives. The attending chores will soon include chaperoning plaintiffs and probably hosting extra children, with or without their families. The pure logistics of it will also include planning for Zealand's days off school and ferrying him to and from the court on the way to work. Already the support rallies are gearing up, and there is such a momentum, such an energy around what is about to take place, that Zealand's parents are learning about much of it on Facebook like everyone else. The excitement is palpable, decentralized, external to much of what Our Children's Trust and its staff might do.

Michael never imagined the case taking on this kind of dimension. "At the beginning I didn't even fully appreciate the size of it," he says; he only vaguely daydreamed that the case could become what it has. Now that it has taken on a life of its own, all the usual activity—family and school and the plaintiffs generally being slammed before court—is rolling up with talk about impending depositions and trying to arrange the holidays around the court calendar. In the bright kitchen of their two-story ranch home, as Michael and Kim discuss these details, they wear expressions that seem tinged with a combination of surprise, bemusement, and a little bit of panic.

This is the backdrop against which I have dinner with Avery McRae and her parents, where we eat at a picnic table in their backyard, under the alder tree that grows through the middle of their deck. Avery is the only child of Matt and Holly McRae, the trio snug in a house of 600 square feet with a player piano and a wood stove and the rest of their life lived outdoors: a kitchen garden, fruit trees, and space for Avery's five chickens and two rabbits. If a plaintiff could be an avatar, Avery would be the animal empath, a kid who started raising money for endangered animals in the first grade and never stopped.

The chickens are named for flowers. Poppy, Lupine, Rosie, Jasmine, and Yarrow. The five are one short of the limit set for backyard chickens in Eugene. Before these, there were others. Buttercup. Dahlia. Iris. Clover. Clover was a rooster, actually, and was retired to a farm.

"Do you know what imprinting is in birds?" Matt asks me. He says this as the legging-clad Avery, flowers trailing down the sleeves of her sweatshirt, comes dashing to the table with Rosie in her arms.

Rosie, Matt explains, "definitely thinks that Avery is her mother."

Avery straddles the bench beside me, hugging and nuzzling the bird, her light-brown hair falling around the bird's face. I look closely and learn it is possible for a chicken to be delighted. Rosie—rusty red with fierce-looking golden claws—tucks her head into Avery's neck and closes her eyes. Avery strokes her wing, holds her tight, and Rosie clucks. It's a chirping, happy sound, one that moves from short bursts to lengthening warbles, like a child's squeal. There is no arguing the bond. Sometimes, Avery says, she calls the bird's name when she comes home, or variations of it, like Tosie, and Rosie runs to her. Now, the comb on Rosie's head—stoplight red—folds into Avery's chin. Primitive. Like the fleshy spine of a dinosaur.

"I read an article one time saying you shouldn't let your children handle chickens and kiss them and stuff," says Holly, right before she admits that she lost that battle and laughs.

The bird makes a flapping play for a grape on the table and is sent back to the coop in favor of a rabbit. A few moments ago, Matt told a story from the weekend, when Avery noticed a spider hitchhiking on the side mirror of the family car. She urged Holly to save it before the wind whipped it to its death. Holly tried but couldn't, and Avery was shaken by this. By the death of a single spider.

It is just like this with this child, they say. It was always like this.

It was the book about snow leopards that triggered the activism. Avery was in the first grade, and the story laid bare the impacts of a warming climate, detailing the effects of habitat loss on the leopard. Avery was dismayed. Holly encouraged her to throw a party to raise money for the leopards. An environmental educator, Holly knew the party was a way to do something, to feel better. Avery raised $200 for the Snow Leopard Trust. By the end of third grade, she'd raised another $250 for wolves and $300 for salmon. She learned more about climate change in the fourth grade at a camp run by Our Children's Trust, then testified before the Eugene city council in favor of the climate ordinance. Later Our Children's Trust asked her, like Zealand, if she wanted to join the lawsuit, and she said yes.

When I ask Avery's parents whether allowing their child to sue the federal government was a big decision, they look at each other—blue eyes meet blue eyes—and nod. For a long time. They say they knew it would impact their lives but they didn't know what that would be like, how stressful it would feel three years later, when they're thinking about Halloween and the rest of their lives with an eight- or ten- or twelve-week trial looming. We all feel it then, sitting in the last light of the evening, the fall breeze swaying the leaves overhead: this sense that normal life is about to be upended.

I ask Avery what climate impacts made her want to sue the government, and she says she thinks about the ocean the most. Acidification. Erosion.

"No one wants to see a place that they love go through that. I've grown up going to the beach a lot. It's probably my favorite place in Oregon, just the Oregon coast, and thinking about what impacts will be there in thirty years if we don't turn this around is pretty sad."

It's not just that the beaches she loves will no longer be there. That she won't walk along Strawberry Hill hunting agates, or watch the seals follow her path along the shore. It's that many of the animals she loves won't be there either. That they will die. Marine mammals, the little critters. Everything.

Scientists have known for more than a decade that we are in the middle of the sixth mass extinction on Earth. Within the year, a consortium of 145 experts from fifty countries will crystalize that view, concluding that one million of the eight million species on the planet are at risk of extinction, many within decades. Much of that risk is just fallout from human impacts, with three-quarters of Earth's landscapes "severely altered" since humans began rearranging things, and 80 percent of wetlands lost in the last three hundred years too. Add climate change, and a temperature increase of 2 degrees Celsius will account for a portion of those extinctions in the years ahead. Even species that are projected to survive will see their ranges "profoundly" shrunk.

The oceans Avery loves will endure some of the most profound impacts, simply because oceans absorb much of the extra greenhouse gas in the atmosphere, warming faster than the earth overall. More than half of the world's coral reefs have been lost in the last one hundred and fifty years already. And oceans are also becoming more acidic, with oxygen levels and nutrient flows changing, too, so that a quarter of fish could die by the end of the century. These losses spell downstream impacts for humans in coastal communities, as well as mammals, birds, and amphibians that thrive near coastal wetlands.

"It's not their fault that we're screwing up their whole future. They can't even go to the courts like we are. I feel like I'm kind of talking on behalf of the animals," or other people, she says, kids like her who don't feel like they have a voice.

It's not a stretch to imagine this particular child speaking for animals. To underscore the point after the dinner plates are cleared, Avery begins a tour of the backyard, a long city lot tucked under pine and bisected by a dome of spirea. Behind the dome—it conceals an outdoor shower—there's a robust kitchen garden ready for winter planting, organized in raised beds. Alongside it, a wire aviary runs half the length of the yard, an enclosure that stands about 6 feet tall and nets not only chickens but a grove of small apple trees too. We step through a door between a fence and the chicken coop, a raised wooden structure emblazoned with a Tibetan prayer flag, each panel bearing a chicken in lieu of the Buddha's prayer.

Poppy, a black chicken with puffy feathers, rules the roost, Avery says. She's not a mean bird. She is charitable to her subordinates, even.

But she's the boss of things. Jasmine is at the other end of the pecking order and—understandably, says Avery—a bit jittery. Avery takes a handful of seeds and tosses them to the birds, then dumps another handful into her palm. Jasmine fusses outside the fray.

"It's hard to get them to her, but if you try, she appreciates it," she says, giving me a handful of seeds. I offer them to the bird, who pecks eagerly enough to drive a hard beak into my palm, grabbing some skin that she tries to rush off with. It's a little uncomfortable, but Avery seems not to notice.

Rosie, she says, loves to sit in the crook of the apple tree nearest the coop. She flies up to roost there most days, about 3 feet off the ground, and while she's content once she lands, she tends to screech on the ascent. Avery demonstrates by flapping her arms, making her best nervous chicken sound.

Avery had just graduated from the fifth grade when the lawsuit began. By then, her preoccupation with animals was in full force.

"She came out with a love for animals that only gets stronger as time goes on," Matt says. "It's remarkable. Her depth of empathy is herculean."

Certain moments stand out. Like the time her parents took her to watch the salmon spawn on Whittaker Creek, and she lay on her stomach on a little peninsula, mesmerized, for about an hour. Or the time—age four or five—she was riding on her great-aunt's Appaloosa, and both got tired so she lay down and took a nap on the back of the horse. There's a picture of it. Now Avery takes riding lessons at a nonprofit where she works at camps or handles therapy horses in exchange for lessons.

We are on the deck watching Matt pull agates from a rock tumbler when Holly mentions the horses. Avery makes a dash for her phone so I can see the pictures, and sits down on the bench again.

Easy Dancer is her favorite horse, she says. She's a paint mix, brown and white, eighteen years old. And she's a horse that likes to move. Avery shows me a video of her riding Easy Dancer at a canter. Both are steady, confident. It's plain to see that Avery likes to run as much as the horse. She tells me several times how cute Easy Dancer's face is. There are a few photos of this, of Easy Dancer's nose pressed against the frame. I see the same face over and over, but Avery sees something else. Expressions and moods that she narrates as she goes.

Then there's Moon Time, who's not as fast but still pretty fast, and good for a lot of different levels of riders. She's a paint too. Then there's Kitty, an older white horse who has arthritis in her shoulder but can still carry the younger kids. Sky Cat and Tabs are the two stallions. Tabs is kind of a pistol but also really cute. Then there's Shadow.

Her phone is full of these pictures the way some phones are full of selfies. She says horse politics are kind of confusing. Because they don't have a pecking order like chickens, they're not consistent in how they behave, and she's still trying to catch on to their cliques and their temperaments. Still, they seem to be in her blood.

Avery has spent a few days each summer riding horses at Matt's aunt's home since she was young, Matt says. He tells me this while rinsing the agates. He and Holly plan to pass the rocks out on Halloween instead of candy. Last year they gave out 350 agates before 7:30, Holly says. This year they hoped to close the street, but the trial will make it too tough to manage the details. Again, there is that sense of foreboding, of time closing in. The music cueing up, the act coming to an end.

Avery tells me she has already ordered the horse onesie she plans to wear for Halloween. Then she takes a brief intermission by climbing a tree. Next, she takes me to the rabbit hutch.

The hutch itself is a wooden cabinet bisected by chicken wire. Each side has a bed, a blanket, a bottle of water, and some food in a bowl. The rabbits can't be together—they're both male and they fight, Avery says—but they are cozy side by side like this. Then she brings me around the hutch. Behind it, ensconced in chicken wire, is a full-fledged warren the size of a four-door car and half as tall. It is quite unlike your standard domestic rabbit habitat. She says her two rabbits made this hilly sculpture after the humans dumped a mound of dirt in their runs and let them dig. Ginger, the red one, is the better digger, she observes, having made an elaborate tunnel on his side of the run. She points out the entrances and exits, and she seems to know from watching how the tunnels flow in between. She traces the routes with a finger through the air. Odin, who is gray and white, over-renovates, she says. He digs until his tunnels collapse. She laughs.

Avery says sometimes the rabbits sleep in her bed, but only naps. She's tried to bring a couple of the chickens indoors but they poop, so it's a no-go.

She takes me to her bedroom—it's getting dark—to show me the spa-like den of fluffy things that is her bed. It's a bit like the galley of a boat. Because their home is tiny, and her family's property is mostly yard (like a lot of urban Oregon homes, an inversion of suburban values), her bed is built on a carpeted platform, its underside a burrow for storage that's not unlike the warren in the yard, only clean and very tidy. The bed is decorated with a deer pillow and the odd stuffed animal. On the wall at its foot is a string of photos draped in white lights. She narrates each picture, moving left to right. There are girls from school, the details of who is still friends with who, how tall her cousins have gotten, and how old her baby cousin is now. She tells stories of trips. Of birthdays. Of shared events. Every third photo, though, is not a person. It is a horse or a rabbit or a cat.

If a child's room can be like a child's heart turned inside out, this one says it all. Artwork of a blue jay, a bear, a leopard, and horses. Tie-dyed pillows and flowers. A button declaring "Rachel Carson Was Right."

A guitar hangs on the wall, and when she retrieves it, the song Avery plays is "A Horse with No Name." Her desk chair is a western saddle mounted on a wheeled crate. On her bookshelf, a favorite book: *Unlikely Loves,* which tells tales of true but unlikely animal romance and friendship. She pages through. The story of the owl and the cat. The fox and the hound. An otter and a badger. Then there are the trios. A rabbit and a guinea pig with a macaw. A turtle and his dogs.

"You know, I love these," Avery says. "It's just these things that bring me so much joy. A tiny cat and farm dogs. Look at that! It's just too much. The seal and the dolphin? And then the goat and the giraffe? They're like best friends! Isn't that cute? Whenever I'm sad, I just read that." She is sitting cross-legged on the carpet, and I believe her. That animal love stories are what she needs when being thirteen is too hard.

It is good that she has this, because being thirteen is about to get difficult. The next time I see Avery, she is on the steps of the federal courthouse in Eugene with her parents and most of the rest of the plaintiffs, looking dashed in a blazer, the *Juliana* trial having been canceled.

There had already been a missed trial date back in February, and Avery had learned from the experience not to get too excited. But this time around, the trial was near enough that the planning seemed legit. Her grandparents were coming for a month to help her attend one day a week and still manage school. There were plans—about who would arrive when, who would be available to do what—and they were down to the granular details.

Then this: "Dad comes into my room and says they have delayed trial. And I pretty much lost my ability to function for like two days," Avery says. It was a Friday. And for the next days, she couldn't find a thing to make her happy. Nothing. She could eat. She could sleep. But she was angry all the time. Angry at the government's attorneys. Angry at herself for falling for this bait and switch. "When you have emotions like that that come quickly and also are super strong, it's really hard to think clearly and do stuff in the well-thought-out way that you were going to. Everything I did was just kind of fumbly and huffy." She says she felt clumsy. Couldn't shake off the mad.

On the steps, Hazel Van Ummersen and Sahara Valentine are standing near her, both fourteen-year-old plaintiffs, also from Eugene. Hazel is sturdy, the smallest of the *Juliana* plaintiffs and a blackbelt in tae kwon do. She's only recently started to drift away from martial arts, fatigued with the competitiveness of fighting, already doing another kind of battle with her first year in high school before this latest with the government. Sahara, whose parents are longtime friends of Avery's, is a hiker and a biker, avid in her outdoorsiness in spite of asthma and also a first-year high school student. Later she'll describe how, in these days after the canceled trial, when school presses on like normal and classes are still the same but the world feels nevertheless off-kilter, she will look for Hazel in the halls, look for Zealand, try to find a connection to the other people who are living this thing with her. Federal litigants. Canceled trial. High school freshmen.

The legal details of what had transpired were dizzying. The briefest way to say it is that Alex was right: the government used all its tools. After Kavanaugh was confirmed, its attorney tried again to stop the case, petitioning the Supreme Court and the Ninth Circuit Court of

Appeals once more to dismiss it. The Department of Justice argued the plaintiffs' claims were overbroad and that the District Court of Oregon lacked jurisdiction to conduct an expansive review of executive branch policy. It also argued there is no right to "a climate system capable of sustaining human life."

The Supreme Court had ruled against a similar effort in a unanimous, single-paragraph decision in July. But when the Department of Justice asked it to dismiss the case for a second time, and the Ninth Circuit to delay trial while the Supreme Court decided what to do, this time the Supreme Court stayed the case, freezing *Juliana v. United States* to consider the government's charge that defending against the plaintiffs was a "burden" and that their claim they had a right to a habitable planet was "manifestly wrong" to begin with.

The legal gymnastics were exhausting, even for the press. Practiced as I was, I wanted a glossary and a running scoresheet, like at a sporting event, so I could keep track of how many such emergencies were normal as opposed to the count we had. Everywhere was befuddlement. Befuddlement and protest. While demonstrators rallied in more than seventy places nationwide, and in four countries, the support didn't buoy moods for long.

By the time Halloween arrives, the McRae family is worn out. By then most of the out-of-town plaintiffs have been sent home—Jayden to Louisiana, Nick to Colorado, other plaintiffs elsewhere, and the college-age plaintiffs back to class—and the fever of the rallies has died down. In its place is worry, depression. Nobody knows what to do with it, including me. After hours of watching Netflix with Larry the cat and the raccoons, my suitcase full of court clothes and groceries scattered, I consider going home to Portland but go into town instead to visit with local plaintiffs and try to get a read on how long the case might be delayed. Plus, Halloween is the stuff of legend in Eugene. And everybody needs the pick-me-up.

The Friendly neighborhood, where Avery lives, delivers on its Halloween cred. It is mobbed with trick-or-treaters trailing parents, or the other way around. At the McRae house, fifteen jack-o'-lanterns stand in a pair of grinning towers and candlelit mason jars burn in the trees. Holly is dressed as a witch, passing out the agates in labeled bags and dancing to "The Time Warp" again and again in the front yard. Kids funnel around the pumpkins and into the grass, drawn to all the other kids already there. Matt kicks up the music with each new mass arrival while Holly gathers the children around her, transformed.

Practiced educator that she is, Holly's voice has a tone that quickly overrides the command codes for anyone shorter than 4 feet. There is a kid dressed as an Uno card and a sumo wrestler in an inflatable fat suit. Tween girls—wearing light makeup and wizard outfits—and tiny satin princesses and little boys with bird wings cover the lawn. A ninja turtle too. All of them set their candy down in the leaves, surprisingly without anxiety, and throw their hands in the air and channel *The Rocky Horror Picture Show* on command. It happens over and over, for hours, for hundreds of people, and Holly is a woman bewitched, a witch bewitched. After two difficult weeks, she wants this, she says. To just dance and make kids happy.

Avery is not happy. Standing in the driveway in her horse onesie, she is aghast. She is thirteen, after all, the age to be embarrassed by your parents. Plus, Avery's funk is in full force. The horse onesie isn't all she had hoped for. The ears are small enough that people keep asking her if she is a squirrel, and the tail is more like a cat's than a horse's. She has her grandmother's dog and people are drinking beer in her yard, and Avery is at home with the flannel and the beards, the mild adult misbehavior and general mayhem, but her humor is faltering when it comes to the rocky horror of her front lawn. Straddling the line between able partygoer and mortified adolescent, her mood an irregular thing since the trial that wasn't, she says hello to her grandmother, returns the dog, and takes a seat on the front steps for what looks, in turns, like seated resignation and mopery.

Two mannequins make battery-operated movements in the driveway, to the occasional surprise of the real people who encounter them. Matt offers me a beer and tells me to read the expert reports in the *Juliana* case. If I can understand those, he says, I can understand everything that is at stake.

CHAPTER SIX

The Furious Paper War and the Direction of the Nation

The government never intends to go to trial. That is its chief defense against *Juliana*: to end it before it ever moves from the courthouse steps into an actual trial court. When I get the rare opportunity to ask its lead attorney why one day, as he walks briskly from a courtroom to an elevator, he reiterates what spectators from the legal community have been saying all along: "The constitutional design has to be respected." In other words, *Juliana* is asking the courts to go too far in telling the government what to do. If the government isn't doing enough about climate change, he says, "that's an argument to make to Congress."

But despite the government's firm position that *Juliana* should be dismissed altogether, the Department of Justice has to prepare for trial anyway, just in case.

Which means that in the days before the case would have gone to trial, before it was stayed by the Supreme Court, the court docket was absolutely stuffed with reports from the experts who planned to turn up and testify. By the time the trial was canceled, everything that might have been presented in a courtroom in the US District Court of Oregon was already there to see—in hundreds of pages written by doctors, scientists, economists, and policy analysts, and in expert testimony on both sides of the *Juliana* case.

This tome—a reading of which, done carefully, takes five weeks—leaves very little to the imagination about what would have happened inside the courtroom had the trial actually occurred, what might still happen someday if the plaintiffs ever get their trial.

This is what it looks like.

In the fall of 2018, the *Juliana* twenty-one were headed into court with an entourage of some of the top climate scientists in the world, not to mention historians, engineers, even a past government appointee with unique insight into prior government efforts to avert global warming. All were offering pro bono testimony in defense of the youth and the planet. And to prove, perhaps most critically, not just that climate change is happening and that it stands to affect the *Juliana* twenty-one but also that climate impacts are already harming the plaintiffs, not just causing them possible harm in some apocalyptic future.

The Department of Justice, meanwhile, planned a defense strategy straight out of the Big Tobacco playbook. Its two parts can be summarized in a few sentences. First, the defense aimed to show that even if the plaintiffs had worsening asthma and allergies from all the wildfire smoke and drought-related dust and increasing allergens, they couldn't prove that climate change had caused their symptoms. And second, while climate change is real and humans caused it, nobody knows when the seas will rise or when the ice caps will melt, so it's impossible to argue that these things could directly harm *these* plaintiffs, since they haven't actually happened yet. Also, if this havoc does unfold during their lifetimes, isn't it already too late to do anything about it?

This second argument was the eyepopper that seemed to be skirting public view. Without the daily limelight a trial would have brought—media circus on the courthouse steps—the broader public was not spotting the fact that while the Trump administration and the president himself were nearly daily disputing the reality of climate change, the government's experts did not, in fact had filed extensive reports asserting that climate change is real and man-made, and that climate breakdown now poses a very real threat to all Americans. At one point, the government's attorneys even asked the court to do away with the part of the trial in which climate scientists would testify. They sought a motion to exclude them on the basis that everyone already agreed climate change is real, so there would be no reason to trot the science out in court.

It didn't work out that way. By the time the trial was set, the finer points about when the seas would rise and the ice caps would melt, and to what extent that would equal specific horrors for the plaintiffs, were still in dispute, and in the end the dispute kept the scientists on the plaintiffs' witness list ready to give the court and the public a full picture of how climate breakdown was already unfolding. The court wanted to hear, needed to hear, to what extent *these* plaintiffs were already colliding with—and would continue to collide with—the distinct fates science portended. Otherwise it would never be clear to what extent these outcomes were the government's fault.

In the twilight of the Obama days, the government had already admitted a lot of the underlying facts. "They admitted they knew about climate change and subsidized fossil fuels," said Ann Carlson, co-director of the Emmett Institute on Climate Change and the Environment at the UCLA School of Law. Among the questions that remained, she said, was "Do you need additional evidence about all the individual ways the government subsidizes fossil fuels?"

Permits. Financial aid. And the authorization of industry practices: air emissions, pipelines, spill cleanups. Nobody really knew what the government—and, in effect, industry—might have to disclose. But she imagined vast evidentiary battles loomed.

Already the fossil fuel industry had funneled billions of dollars into exactly the kind of public disinformation campaigns that once caused people to embrace cigarettes as a symbol of freedom and personal choice, even while science showed cigarettes were killing them. (More later on industry's forays into disinformation and doubt.) Chevron's attorney, Ted Boutrous, speaking on behalf of five fossil fuel companies in a different climate case in San Francisco, told the court that fossil fuel *demand* was the cause of climate change, not *supply*, which is a whole lot like saying that even though cigarettes give you cancer, and cancer kills people, smokers assumed these risks when they could have just listened to the science. The fact that such science had been obscured by industry since the tobacco companies first invented that move was, at least for the sake of Boutrous's arguments, suddenly immaterial.

Considering the relative success of this strategy, there was no reason the government should have avoided it. At least if the objective was to win. Being long past the ethical yardsticks that would have kept the nation out of this arena in the first place, the tobacco defense led to good outcomes—read: financial windfall—for that industry. Despite the fact that people think tobacco companies lost in the end, did they really?

Referring to their $206 billion settlement with states over the public health costs of smoking in 1998, Michael Liebreich, a former advisor to Shell New Energies, summed it up in a column for BloombergNEF as more business decision than defeat. "The companies calculated that it was worth doing a deal in order to remain in the cigarette business and they are still highly profitable today," he wrote. Now it's the sale of cigarettes that funds continuing payouts to the state governments that sued over the public health costs of smoking. Hardly the template for climate reparations, especially if you consider that funding the fix necessarily perpetuates the problem. Also, not a coup for the public where health and safety are concerned. Between the 1950s, when the tobacco companies first rolled out this legal defense, and its monstrous settlement in 1998, the industry acquired a whole new generation of smokers and another forty-plus years of profits. And they're still in business.

Every day of delay in halting climate change similarly promises more supply, more use, more profit, and future business for fossil fuel companies. And there could be no better path to assuring industry's course in delay than for the federal government to appoint an industry attorney to defend it in the *Juliana* case. Which is exactly what happened, though it is perhaps the strangest part of this story.

Jeffrey Bossert Clark had spent years settling claims for the Deepwater Horizon oil spill on behalf of BP before Trump nominated him to lead environment and natural resource litigation for the Department of Justice in June 2017. Clark had served in the division previously, under George W. Bush. And in this way, he'd spent much of his career in the revolving door of employment between the Department of Justice and Kirkland & Ellis, the private law firm that represented industry in some of its highest-stakes litigation. At Kirkland & Ellis, where Clark became a partner, he figured prominently in industry efforts to ward off government regulation of greenhouse gas emissions in the Obama days. Today, Clark supervises "all federal civil environmental and natural resources litigation involving agencies of the United States" and personally handles "select, high-profile cases," including *Juliana v. United States.*

The optics of having a former BP attorney leading the government's defense was certainly symptomatic of the larger problem: the government failing to separate its own interests from the interests of industry in *Juliana* and a lot of other matters. But nobody seemed worried about optics, at least not openly. Clark's nomination coincided with the fossil fuel industry's withdrawal from the *Juliana* case, though industry had fought hard to intervene under Obama. Having an entire industry step out of such consequential litigation was partly due to the pesky fact that some of the biggest oil companies in the world were about to be subject to discovery and probably not so keen on turning over documents. After all, they were being sued for securities fraud by stakeholders, in nuisance cases, in product liability cases, and by others seeking climate redress. If they turned over documents in one case, they would likely have to do so in these others. And who knows what criminal charges might surface for individual actors. That aside, the industry also didn't need to represent itself anymore. Its lawyer was already in the ring, helping the government do it for them.

Clark was just the man for this job. He'd been hostile to attempts to regulate greenhouse gases for enough years that he was well versed in how to dismantle these regulations. In 2010, Clark had given a talk at the Federalist Society's National Lawyers Convention called "EPA: An Agency Gone Wild or Just Doing Its Job?" In it he'd characterized US efforts to regulate greenhouse gases as akin to a socialist power grab for the top tiers of the economy. This just a few months after Clark authored a column on the right-wing PJ Media blog that asked the rhetorical question: When did America risk coming to be ruled by foreign scientists and apparatchiks at the United Nations? His answer: the moment Lisa Jackson, the EPA director under Obama, issued a rule determining greenhouse gases endanger public health and welfare, resting largely on UN science. "The United States is not a technocracy, let alone a UN-ocracy. It is a republic—but only for as long as we can keep it," he wrote.

"Here's what plagues me," said Mary Christina Wood. "The big question, sort of the elephant-in-the-room question, is why on earth is our government representing fossil fuel interests when doing so is going to bring civilization to an end? If they succeed, their success would drive us over the climate cliff. How did we get here where our government is on this tyrannical track that would lead, literally, to massive death, destruction, property loss, misery, economic loss? How did we get here?"

That part isn't as clear. But what is clear is that the government planned its own expert witnesses to make the bedrock case against a climate remediation plan. None of its witnesses were climate scientists. Instead, the government planned to have three medical doctors testify that even if some of the plaintiffs were sick—with asthma and allergies and the like, as Nick and Kiran and some of the others claimed—it couldn't be proved that climate change was the culprit. Two professors tied to the Energy Modeling Forum (EMF) at Stanford University, which had long been funded by industry to influence energy policy, would testify about how hard decarbonization is, among other things. An MIT scientist who worked on carbon storage planned to testify about how difficult that is too. Also, a Cato Institute scholar on land management planned to testify on the possible pitfalls of planning for carbon impacts. The government's expert roster also included David Victor, a researcher of climate policy who literally wrote the book about why the world hadn't made progress on a global climate accord. He was the cheapest of all of them at $325 an hour. One of the Stanford professors, James Sweeney, charged $800. Two had served on panels of the Intergovernmental Panel on Climate Change (IPCC), three had served or worked for government, and at least half had ties to fossil fuel companies. They were typically in the employ of either medical practices or universities. Now they were also in the employ of the public to defend the government.

Small wonder the Trump administration ducked this trial. What loomed looked like an expensive boondoggle, full of opportunity for questions about ethics, character, conflicts of interest, and the direction of the nation and its learning institutions. And since you don't get in the ring with a Million Dollar Baby, *Juliana* fit the bill for the next best legal tactics: dismissal and delay.

The case was always going to take years, even without the government's emergency petitions and the consequent slowdowns. If the District Court of Oregon decided it quickly, it would face another three years of appeals, at least. The Ninth Circuit Court of Appeals first. Then, likely, the US Supreme Court. Many of the *Juliana* plaintiffs would be college graduates by then, most high school graduates, at least. The outcome was a long way off. But the trial—the trial was also for a nation that needed the spectacle, the epic, the television version of the conversation it needed to have. If not over fires and floods, the prospect of the impending end to us all, there was this stuff, so American: made for TV.

Who could help but root for a bunch of young people against a primetime cast of villains, anyway? The social cachet the youth would have—zoomers, all of them—in a court stacked by and for baby boomers and a few Gen Xers could have catapulted the plaintiffs to infamy, if not spurred an intergenerational culture war that could have spilled into the streets.

But the fear that the government's case would have stoked in Americans? The great swaths of the nation where people worried over basics like affording bread and milk while the prescriptions and the rent were breaking them? That fear was for real. Even though 71 percent of Americans believed in climate change in the fall of 2018, only 57 percent were willing to pay to help fix it, and even then they would only pay $1 a month. Faced with a fee of $40 a month to halt a climate breakdown, public support plunged by more than half, to 23 percent. Americans were similarly lukewarm about a carbon tax. In other words, people knew there was a problem and wanted the government to fix it, but they were very worried about what it would cost. The debate about remedies brought forth this undercurrent of unease people already had about whether reacting to climate breakdown spelled financial chaos.

And then there was this: one study found that climate denial in America also stemmed from "the strong ideological commitment" to personal freedom and a lack of government regulation by small-government conservatives and libertarians. "US climate deniers often rest their case on the defense of the American way of life, defined by high consumption and ever-expanding material prosperity," its author found. Many people just did not want to change. And the study found there was a lack of forthrightness, of understanding even, on the part of the proponents of change about just how much American life would really have to.

These distressing questions of how much it would cost to avert a climate catastrophe, who would pay, and how much American culture could be altered by it, they were the questions the government's experts were poised to raise in the form of critique.

The *Juliana* experts planned to argue past these fears. Or to try. Mark Jacobson, a professor of civil and environmental engineering at Stanford University and director of its Atmosphere/Energy Program, offered dozens of pages of documents arguing it was possible to convert the US energy system to 80 percent renewable energy by 2030, 100 percent by 2050. He said it could be done using mostly existing technologies—things like solar, wind, and wave power, hydroelectric and geothermal. And he wrote that such a transition could create two million jobs and spur economic gains, all while minimizing harm from air pollution and climate change and energy insecurity, something he had modeled using computers. "Based on the scientific results presented, current barriers to implementing (wind, water, and solar) roadmaps are neither technical nor economic. They are social and political," he wrote.

James H. Williams, a professor of energy systems management at the University of San Francisco, similarly said in his expert report that it would be possible to deeply decarbonize the atmosphere to 80 percent of 1990 levels by 2050 without everyone in the nation having to be on some sort of energy rations. What he meant by *energy rations* was, ironically, the kind of brownouts that California would soon be facing to avoid wildfire—brownouts that might also be necessary if the world waited too long to decarbonize. Williams added, though, that these measures might not be enough. And if the US wanted to avert the catastrophic impacts of climate change by decarbonizing another 20 percent, it could. People might have to spend more money, and "it will likely require some early retirements of fossil fuel infrastructure"—things like coal plants, pipelines, refineries—but it could be done without drastic reductions in the amount of energy our consumption-hungry nation used day to day. In the end, it would "not diminish basic quality of life and standards of living" in America, the very things people feared.

In the furious paper war that ensued, however, Howard Herzog, a researcher in the MIT Energy Initiative, attacked these ideas on the government's behalf, raising questions about the ripple effect of the transition in the marketplace, the unknowns. Things like lag times in siting new energy facilities, in getting permits, things that could add unplanned cost. And he said Jacobson "fails to comment on the numerous coordination issues associated with rapid large-scale transformation of the US energy system." Herzog suggested people might not be willing to charge their electric vehicles during the day, for example, something a renewable energy system would require. And that dams might not be able to spill the needed amounts of water for hydroelectric at times because of competing needs for irrigation and water for fish. Generally, the picture he painted was of an America that would be lumbering through the transition in a staid, rigid bureaucracy, even as the planet was in a state of emergency. The questions he raised were often good ones. But the answers were unknowable. They could only be answered by trying, by living in the lab. To stir the fear that it would all be harder than anybody said and cost more, too, was to suggest that things would be better if we left them alone. As if it were better to just die on the operating table than *do* the surgery.

This argument from a government that landed the first man on the moon and invented the Internet, in a culture that built the first personal computer, cell phones, telegraphs—it seemed absurd. Suddenly this nation, the one with the world's most powerful military, the one whose moral compass and courage won the Second World War, was cowering. Afraid of risk.

It was hard to fathom what America really stood for in such a moment.

James Sweeney, who analyzes energy economics and policy for Stanford's EMF, painted a dystopian reality at the point when the nation would no longer permit fossil fuel facilities. Prices would fall. The rest of the world would just use more fossil fuel anyway. The US would have to import fuels. Suddenly national security would be at risk. Consumers would have to change behavior. And leaps in technology would be required too. The energy systems proposed by the plaintiffs' experts "could become a reality only if the government abandoned free market principles throughout the economy, adopted a command-and-control approach, and mandated the adoption of the technologies proposed in their energy system, or provided massive subsidies, which would require tax increases," he wrote. It was all very full throated. It was also like a very scary movie, except usually in the movie there was a meteor headed for Earth and people were trying to run from the problem, not trying to solve it. The government's experts offered nothing in the way of solutions, only these sorts of defenses of the status quo.

It made me wonder what the cafeterias were like at Stanford, where Sweeney and Jacobson stalked the same ground after submitting expert reports on opposite sides of the *Juliana* case. So, too, at symposia, where experts similarly stood crosswise in the litigation. I imagined fights over the last of the hot dogs, or at least mean looks. Parts of the reports reflected this kind of hostility, the prose so searing, so personal and pedantic—"Erickson demonstrates lack of attention"—that it sometimes read like a grade-school spat for very educated adults.

I asked Jacobson what it was like to share the academic bubble in these times.

"It's not normal to be at odds with people in your own community," he said. "I just basically ignore them and they ignore me. It is very awkward."

It was especially awkward since, around this time, Jacobson's office was three doors down from Sweeney's. A third Stanford professor, John Weyant, was also an expert witness for the government, and served with Jacobson on some of the same university committees.

"I go to one meeting where there are only five people in the meeting: me, Weyant, Sweeney, and two other people." A difficult meeting at which all three tended to communicate in circles.

There had also been a lawsuit, Jacobson charging another scientist and a journal with defamation over a critique of his work by a group of authors that included Sweeney and Weyant and David Victor. Jacobson charged the scientists never actually reviewed his data, just wrote the paper, but he withdrew the lawsuit later, saying it took too much time and money to sue.

Similar disputes among the experts—and deep ideological debates about the ethical role for scientists in defending the government against *Juliana*—were under way in other scientific fields and at other institutions.

Take land management. Climate remediation would call for new practices to decarbonize the atmosphere. After all, you need a certain number of trees for such things, and for soil to be carefully managed. The plaintiffs' expert Philip Robertson—a professor of plant, soil, and microbial sciences at Michigan State University—charted a path to maximum carbon storage and said 20 percent of the global need could be met on US soil, about a third on federally managed land, the rest private, with policy and funding for such measures being the only obstacles. These were big obstacles. Saying that you could control people's land seemed to violate some cardinal rule of American values, except when you considered that the government had been buying these kinds of outcomes from farmers for years. Still, these were the very things conservatives abhorred: restrictions on personal property, on freedom.

Daniel Sumner, the government's expert, said the plaintiffs hadn't proved these investments were reasonable or cost-effective. And his points about what such measures would do to the cost of food, to rural economies, to other conservation goals were cautions that would alarm ordinary Americans. They were perhaps the strongest and most troubling arguments the government had. But they were policy questions, too, questions typically answered through policy making in other situations. Like what people should be made to recycle, prevented from dumping in the form of hazardous waste, and forbidden to do in order to protect rivers and streams. True, some policies would come with big cultural adjustments. Things that in an emergency people would also have to get used to. They were harder than saving string and aluminum foil, the woman taking on jobs in factories, as they had in World War II.

The bigger fright, economist Joseph Stiglitz argued, would be the cost to younger people if the nation did nothing.

"More than half a century ago, President Johnson sent a message to Congress that we faced two paths: the cheaper option, in the short term, of carrying down the path of pollution, or the more expensive option (at the time), of restoring the country and its natural heritage to the people," Stiglitz, who taught at Columbia, wrote. He said for the last fifty years, the government had "shirked from the 'more demanding' course of restoring 'America . . . to her people.'" With policies that discounted the eventual impacts on children, the government continued "to steer America on the path of incalculable losses and away from that more demanding and sane course. The costs of fixing the damage today are much higher than they would have been in 1966 when President Johnson sent his message; but, the costs today are much lower than what they will be after another fifty years of fossil fuel pollution and inaction."

Stiglitz had a Nobel Prize by then, shared another with Al Gore. He had also taught at Yale, Stanford, Princeton, and Oxford. He was the kind of ringer who could make any opposition weak kneed just by stepping into a room. He told the court in his report that if catastrophe hit before the nation could fix things, it would take personal wealth to escape the perils of climate change. That some people would lose homes while the tax burden of disaster relief would skyrocket, right along with the rising costs of insurance, of food, and of health care in a society suddenly rife with diseases like Zika, carried to new lands by mosquitos that would range farther in warmer temperatures. The economy would weaken, increasing income disparity. Some people would have, others not. And everyone would need more money to dig out of the situation at a time when less money would be available.

Stiglitz compared what the plaintiffs were seeking—a full carbon accounting of the nation, plus an order to create a climate remediation plan—to the simple practice of having business insurance. At a certain point it just made sense for the nation to do this. There wasn't some third-party insurer waiting to make financial bets on who would be harmed by climate change and who would not. No free market insurer wanted to make this wager, which said everything about the degree to which the free market could ever address climate change. And climate change would hurt developing countries much worse than America; the cost of adaptation would be "well beyond anything that those countries can afford." What that meant, Stiglitz said, was because the United States made the bulk of this mess, responsible as it was for 25 percent of historical carbon dioxide emissions across the globe, it would be American youth who might be tasked with cleaning its share someday. If not for their own sake, then because mass migration from other nations would force them to have to.

To look clear-eyed at this, at about fifty pages of frightful prognosis for American youth, was to understand what Greta Thunberg would mean, in another year, when she would tell the UN Climate Action Summit that if governments continued to do nothing about climate change, left all the problem solving to the youth, "We will *never* forgive you." And scientists were seeing, in increments since at least 2014, that they had already greatly underestimated how fast this future was coming.

Lise Van Susteren, an expert on the trauma effects of disaster, drew on literature from the nation's already lengthy list of superstorms and hurricanes—Katrina, Charley, Sandy, bad enough to remember by name—to conclude the *Juliana* plaintiffs and lots of other youth were being harmed by climate change in ways that would lead to lasting psychological impacts. And even while the government's expert discounted their duress, since none of the plaintiffs were suffering from acute mental illness, a third expert likened this dismissal to telling the kids from Flint, Michigan, that even though there was lead in the water, if it hadn't made them sick yet, they should just keep drinking it.

The tall task of drawing out this joust in front of a judicial panel would have been up to the attorneys. They had twelve weeks to craft the trial. Climate change as theater. As public performance. As education. In addition to the expert witnesses planned, they also had others to support basic facts, including people who had previously worked for the government. To craft this spectacle on the courtroom stage would have been a tense dance between fact and fear, played out in the slow theatrics of legalese.

Still, it would have been an epic show. To be in this room, to witness the trial for humankind, the plaintiffs in the galley, all buoyed by the relief and stress of finally having made it there, it was a reporter's dream—like Scopes, the Monkey Trial for the climate. Though no chimpanzees would dance at the courthouse, the spectacle of *Juliana v. United States* was already primed. Daily press conferences had been planned. Major media was set to attend. I'd agreed to cover for Reuters, plus reporters were being sent by the *New York Times*, the *Washington Post*, and Bloomberg, to name just a few. Other events in the case had been so stacked with press that the courts had reserved rows of seats just for the journalists that would arrive in droves to cram in.

Advocacy groups, lawyers, civic organizations, and others also had a stake in the case, and having papered the courts with friend briefs on behalf of the plaintiffs, they were also primed to dive in. For months, the *Juliana* press team had been grooming their hashtag—#YouthvGov—and the attending social media was expected to trend big. If past retweets of case updates from Ellen DeGeneres and Leonardo DiCaprio were any barometer, it would have. The documentary crew was still in tow, having followed the plaintiffs in the three-year lead-up to trial and launched their Kickstarter campaign for production costs in September, riding the pretrial wave of excitement. By early October, there'd been so much buzz that court staff was telling anyone who wanted a seat to arrive at 6:30 a.m. and prepare to stand in line, like for a rock concert, for the 9 a.m. trial time.

The trial would have drawn the media eye to climate change and held it for weeks, months, a fact that could have changed the national conversation about global warming for good, win or lose. It was a moment I did not want to miss. And when I missed it, when we all missed it, what it felt like was the nation's loss, a missed opportunity to join the civilized world in a conversation most other nations were already having: How do we fix this? How does humankind survive?

Attorneys were expected to make opening arguments with a full complement of heartstrings: Jayden sleeping on her living room floor with her single mom and siblings. The fires plaguing western states, forcing Nick to spend his summer indoors listening to the air purifiers. The trial was to be a narrative, and each of the *Juliana* plaintiffs was to tell, through experts and exhibits, the story of the warming planet. Each had a unique role to play: that of ambassador to the crisis, chronicler of how it looked through the lens of their youth. They were uniquely able to speak about what the breakdown of the planet's ecosystems meant for them and for other young people their age. And for each of them, climate change was personal.

To strike the balance between this emotional tenor and the terrifying, sometimes stultifying facts of the problem, renowned climate scientist James Hansen was to have taken the lead, stalking the line between fact and fear. Ever since his landmark 1988 testimony to Congress in which, as a NASA scientist, he told the government that climate change was real and human caused, and that the time for action was now, he'd had a unique ability to connect with audiences and hold their gaze. He'd commanded stadium crowds, front pages of newspapers. Thirty years later, his granddaughter, Sophie Kivlehan, a *Juliana* plaintiff, had spent a lifetime watching him do it.

When she was young, she said, he would bring her to his speeches. Speeches that were televised, one audience so large it filled a football field. At interviews, people would ask her questions too. "He would bring me for kind of that visualization . . . needing someone younger than you to make you feel that emotional motivation to do something." This compulsion to act was a thing Sophie, at twenty, now felt when she looked at Levi, the youngest plaintiff in the *Juliana* case. Hansen knew the impulse well, could play to it, align the climate science with this basic instinct to protect the young.

"He's very matter-of-fact. And he does his science in his office and presents it, and at the end and at the beginning he will say something emotional. And throughout it he will say something emotional to get people involved," Sophie said. "But at some point, he defined this balance of connecting with the emotional burden of it and disconnecting with it so that he doesn't burn out."

Hansen could do that for listeners too. Show them the child and let them turn back to the facts, countermand the horror of what was at stake with glimpses of what could be done about it. He'd kept in the fight longer than most this way. And it was how he taught Sophie to integrate climate advocacy into her life. How to separate herself from the movement and not exhaust herself like so many advocates she had known, attending every march, no time for jobs. Now Hansen would do this in the courtroom too. Try to inspire both the courts and the public to act without causing them to turn away.

When he was through, the *Juliana* attorneys planned to move through the effects of climate change piece by piece, letting their experts walk through its dimensions. Warming oceans, rising seas, superstorms, worsening wildfires and droughts, glacial melt, collapsing ice sheets, intensifying rains, floods, the collapse of ecosystems, species extinction, biodiversity loss, ocean acidification, dying coral reefs and marine life, a collapsing food web. In between, each plaintiff was to tell their story. Stories of bedrooms that filled with water in the night, of beaches set to be underwater, and of centuries-old family farms threatened by encroaching wildfire and drought. Stories of asthma made worse in summer heat, of forests destroyed by bugs that no longer died for lack of cold. There were ice storms, sinkholes, wells run dry, whole populations relocated from rising seas, melting glacier parks, and beaches lined with dead fish.

No one knows what these stories would have looked like in a courtroom. But as November 2018 begins, it is up to the Supreme Court to decide if there will ever be a trial that could tell them all.

CHAPTER SEVEN

Soon to Be Disrupted by Wind and Rain

These stories, they are especially horrid when told by children. And many of the *Juliana* plaintiffs, three years into their court battle, are very much still children. Levi especially so. After trial is canceled, he flies home on Halloween, dressed in a Captain Climate costume handmade by his mother. The costume consists of a long-sleeved blue shirt and matching leggings, accessorized with an insignia globe, a cape, and an eye mask. He has a song that goes with it, too, a modified version of the *Captain Planet* theme song: "Captain Climate! He's our hero!" Except in Levi's version fossil fuel emissions get the takedown in the end.

His homecoming is bittersweet. Levi's going-away party was less than two weeks earlier, and he'd been sent off by friends from his church for what was supposed to be an extended stay in Oregon for the trial. Now he is back on the barrier island between the US mainland and the Atlantic that he calls home.

Levi and I met there the same day Hurricane Michael hurtled toward land, October 10, 2018, less than three weeks before the trial had been planned to start. Satellite Beach is just south of Cocoa Beach, epitome of white sands, and I only narrowly succeeded in flying ahead of the hurricane to reach Levi there. I expected to be stuck in Florida until I could get out. And while Hurricane Michael derailed travel from Denver to Charlotte and decimated much in between, Levi and his mom, Leigh-Ann, showed me around as the sun still shone to the east, a condition soon to be disrupted by wind and rain.

Levi is standing outside the car door when I open it. To my utter horror, I am driving a sports car. It's a newer Camaro with awful blind spots, and even though it is considered an upgrade from the standard economy car, it is also what the rental agency gives you at 2:30 a.m. when they don't have anything else. Having flown from Jayden's home in Louisiana, I came through the path of the hurricane as it approached, a circumstance that made for delays and uncertainty on top of the usual Dallas thunderstorms. Thus this car. Levi says he likes it right away, but then he doesn't have to drive it. In addition to the awful blind spots, the defogger doesn't work, a rough discovery on the dark stretch of swampy road between Orlando and Levi's home. Now it's afternoon, and some of us are still foggy. Hurricane Michael is set to make landfall in the panhandle—on the other side of the state—any minute. It's still dry, but the wind is starting to blow, the odd gust pitching up to 25 miles an hour.

Levi is wearing a striped swimsuit and a button-down shirt patterned in yellow lemons. He greets me with the bald curiosity with which he greets most things: head cocked, big blond Afro blowing, like I am a strange exhibit at the zoo. He starts with questions about the car: "Is it a Camaro? How fast does it go?" Though Levi has been interviewed dozens of times by now and—it turns out—is good at talking about climate change and his case, his natural disposition is also as inquisitor, extrovert. It takes a few minutes to sort out which of us will be interviewed.

His mother explains that this is how Levi learns so much—he is always asking questions. On a recent trip to Calaveras Big Trees State Park in California, she says, he cornered a ranger and grilled the man for an hour. "We're, like, sitting there eating lunch and he's asking the ranger four million questions. I'm like, 'All right, cool. That's our homeschool activity for the day,'" says Leigh-Ann. She is thirty-two, petite and blonde like her son. She tells me this while Levi is walking to and from our post at the beach, interviewing the various other people who turn up and reporting back the details in enthusiastic dispatches.

"He does that kind of thing all the time," she says. "We went to the courthouse on a field trip and him and one other kid—it was like two years ago, they were like nine—the two of them just bombarded these poor people with all these questions. But man, they learned so much. And even I learned a lot from just hearing like, 'Well, what's this? What's that for? How do you figure out this? What happens here?' They even had them in this little room that was the holding cell and they said, 'This is all you get. You get a sandwich and it's a peanut butter and jelly or a ham and cheese.' And Levi goes, 'What if you're allergic to nuts and dairy and you're a vegetarian?' And they were like, 'Well, too bad, you eat the apple.'" He got a taser demo too.

But this conversation is later. First, we walk onto the beach—a thin white strand fringed with seaweed—so I can get a better understanding of Levi's claims against the government. He lives on just 3 square miles of land between the Atlantic Ocean and the Indian River Lagoon system, the saltwater inlet that runs between the island and the mainland. And once on the beach, Levi is quickly off to the surf, thrilled with the viciousness of the tide as the weather begins its turn. "Look Mom, there's perfect waves!" he hollers.

His homeschool group adopted a nearby beach for cleanup, so Levi is diligent about litter patrol even now. As we walk along the water, he has a plastic bag from a nearby dispenser that he starts filling with bits of trash. Bottle caps. Little scraps of whatever. A tiny rope he briefly turns into nunchucks. While he does this, we talk some. We start with how fortunate it is that Hurricane Michael will not make landfall here. Because of climate change, the annual tropical storms and hurricanes have increased in intensity in the last twenty years. And storm by storm, the changing weather is taking a toll.

"That's the biggest problem for me and my family," Levi says. He's been evacuated from the island three times in his life. "When you leave the barrier island, you don't know what you're going to come home to or if you'll even have a home when you come back."

Scientists are still learning what impacts climate change will have on the frequency of hurricanes. But they've determined that when hurricanes do arrive, global warming likely boosts their rainfall and wind intensity. That's also true for tropical cyclones—those storms will become increasingly vicious on a warmer planet, escalating to Category 4 or 5 storms more often, with rainfall up to 15 percent heavier. All of this boosts the odds Levi will be exposed to awful weather in his life, storms that are more than threats to his home and his island, his way of life. They are also threats to his emotional well-being, his sense of place and his identity.

Some of the damage these storms inflict is what you'd expect from severe weather. Hurricane Irma wrecked Levi's school and knocked trees down behind his house in 2017. The trees took down the power lines, too, and cut off power for two weeks. Levi says the storms also cause flooding in town, and though his family sandbagged in 2017, the flood still walked straight up to the front door and waited creepily on the steps.

"I could clearly kayak down the street if I wanted to," he says.

Then there are the other effects. Things that mix with island life as the climate slowly recalibrates. Like the fact that storms overwhelm the sanitary system here, and add bacteria to the seawater that threatens fish and promotes the algae growth that can kill them. Or that storms and sea level rise are damaging the beach. Seas are already rising along the coasts of Florida. The EPA predicts that if oceans and the atmosphere continue to warm, sea level is likely to rise by as much as 4 feet along the Florida coast in the next century. That prediction will double within a year, but it's still too much to sustain civilization on this island. In its resiliency plan, the City of Satellite Beach found that 2 feet of sea level rise would be a tipping point, something that's predicted to happen by 2050 if global warming trends persist. In between, the storm surges, tidal flooding, and the bacteria problem will continue to affect the island in worsening ways. At 2 feet, the island's fire station will flood in some tides, along with nearby streets. At 3 feet, the city's community center, several roads, and the city hall parking lot will be underwater. At 4 feet, city hall itself will be submerged.

Levi says native plants have been put in the dunes to help prevent erosion as water roils them in storms. He points out a trio of plants—sea grape and grasses—and says they strengthen the shore. Then he picks up a seed and shows it to me. It's called a hamburger seed, and it looks like a burger in a bun. They're one type of the many buoyant seeds that drop from tropical trees and vines here, many of the plants legumes. These hard-shelled seeds have air pockets and float on the water until they wash ashore, seed a plant somewhere far from where they first dropped. They can travel a long way on Atlantic currents—anywhere, really; Jamaica, the United Kingdom—which makes collecting them fun. Levi can identify many of them and is surprised when I don't know what a sea bean is. He adds my sea bean education to the day's tasks.

Beyond the sleeves of his blazing yellow shirt—also sewn by Leigh-Ann, who is a master seamstress—Levi's skin is covered in the scars of youth: bug bites and scrapes and a hefty pink scab that marks his right elbow. Signs of an active life. And as we talk, his energy is obvious in the way he takes breaks to buzz around the beach, then comes back with things to show me. A piece of sea glass. A shell. Another sea bean from a Mucuna vine. He knows the names of the shells too. When he finds something he isn't familiar with, Leigh-Ann says, he takes it home and tries to look it up in a reference book that he likes. He reads fiction too. Percy Jackson. Harry Potter. It annoys him that his co-plaintiffs talk about movies he hasn't seen.

Being the youngest? "It stinks," he says. "I like it because all of them are like older siblings to me and they're all really nice. But in other ways, I don't know. Sometimes . . . " he trails off.

Leigh-Ann jumps in. This is normal with them, Levi starting the talking and Leigh-Ann finishing.

"They talk about teenager stuff and he's like, 'Huh?'" she says.

"Yeah," he concedes. "Conversations a lot of the time are boring." Then he looks at me squarely, nearly wincing, and adds, "They had this argument about whether or not a water molecule was *wet*." He says this, incredulous. As if there is no greater torture than this kind of talk. He doesn't know what he'd rather talk about. But it isn't this. Or legal talk. Later he'll say that court talk is boring too.

He turns back to picking up trash, digging, running into the waves, then jogging back again, peppering the in-between with fresh details about hamburger seeds and grass. At the water's edge, he calls out, "Can I get my shirt wet?" and I learn later this is the kind of question he asks when he knows the answer is no. He is a boy like I remember boys being before we started putting them in front of screens: dynamic. And as I learn more about him, I learn he leads a screen-free life, a kid unto nature. Which makes it easy to see, in his everyday play, what he will lose if this island is wrecked by storms or underwater in fifty years. And those outcomes, they are likely.

"The thing that I worry about the most is the fact that I might not have a future. If I have children, I might not be able to show them where I live," Levi tells me later. "And my greatest fear is that the barrier island that I live on right now will go underwater and I won't be able to go and visit the place where I grew up and where I've lived for most of my life."

Only coordinated worldwide action to reverse warming trends, or at least action by the world's biggest polluters, will stop this outcome. Human-induced climate change has been causing Greenland and Antarctic ice to thaw since the 1990s, when the warming atmosphere transferred heat to oceans to the extent that melting began. All this ice melt, combined with the fact that warmer water expands, is making sea levels rise steadily. If it hasn't already, this process can spin far out of human control. Already, seas are rising faster than climate models predicted, largely because those models didn't account for feedback loops that accelerate such things. Like the fact that oceans will hold heat for thousands of years and continue to melt polar ice, even if the world reduces greenhouse gas in the atmosphere.

Other measures are only temporary. The city dumps a bunch of sand on this beach to protect it sometimes, for example. The shore used to naturally get big and small, erode and refresh year after year, and the seaweed helped to recover it. Now it only ever gets smaller. Two years ago the city put an artificial reef in the water to attract juvenile sea turtles and tropical fish from a natural reef nearer to shore, one that provides habitat to fish and birds, sea turtles and invertebrates, and also protects the island against waves in storms. The plan is to cover the natural reef with sand to claim more beach. Shift the wildlife to the artificial one.

"It's kind of unrealistic," Leigh-Ann says. The sand is a crowd pleaser. But she points to the high-water line above our heads in the dunes and says the water level is such that the coastline will erode no matter what. "It looks good for a little while." But the city has spent millions on sand, twice, she says, and both times it was carried off within a year. The fact that it isn't native only adds to complications for marine life.

Levi joined the *Juliana* case at the suggestion of a pastor at his Unitarian Universalist church because he is a child who is slowly losing his way of life. Which is why, despite his age, he understands these truths, and talks easily about climate change and what it will do to his world.

It's not a rote speech, a cause he can't feel. He's given many a talk, many a courthouse stump speech, and, more so lately, many an interview. CNN. ABC. *Tampa Bay Times*. Public Radio International. Soon to be NPR and others. But such practice does not curb his anxiety. He says this kind of advocacy makes him feel empowered, like people are listening to him and to what he has to say. But so far none of the reaction to this work is moving fast enough to outpace the destruction of his island. In two weeks, a red tide will cover this beach in dead fish, and Levi will be stuck in his house trying not to breathe the air, barely able to go outside without choking and feeling like he is "inhaling a fistful of powdered black pepper."

As we leave, he shows me a stone sculpture made from coquina rock, a type of sedimentary rock that forms from shells and sand. In a sand pit at the entrance to the beach, sculptor Monty Fein—whom Levi has also debriefed—has shaped these rocks into dolphins, starfish, turtles, a crab, all of which Levi points out to me. There used to be mushrooms, he says, but not anymore.

I ask him what people say to him when they talk to him about his case. He says, "Some people are very optimistic about the lawsuit and others are very supportive. It just really depends on the person. Some people have no idea what it is about or what climate change is or why I'm suing the US government. It just really depends on"—he pauses for a minute while he finds the right words—"their views on climate change and things like that, I guess."

I ask him if he is able to explain why he is a plaintiff to people who don't understand climate change or believe in it. He says, "Yeah. Sometimes I do try. And other times, when I can just kind of tell that it's a hopeless case, I just say, well, okay."

Levi says he doesn't mind that other people have different views. Then he points out a snail in the sculpture, tells me his last interviewer had a really cool drone to photograph it with, and is off in pursuit of a lizard.

As Levi and I were having this conversation, Hurricane Michael made landfall on the other side of the state with winds of 160 miles per hour, delivering exactly the kind of devastation Levi has so far been luckily spared. It hit shore at 1:30 p.m. near Tyndall Air Force Base in the Florida panhandle, ripping the roof off the elementary school and damaging the roofs of every other building on the base. It wrecked the drone runway, a training facility, and a materials research lab. It was the fourth most powerful storm to ever make landfall in the United States, and the most powerful storm to ever hit Florida.

Complicating matters, just one day before it came ashore Hurricane Michael had been rated a weak Category 3, so many people didn't evacuate. Some didn't have the option because they couldn't afford it. That the situation didn't look bad enough to leave—many a weatherman would be wrong about this—made it so a lot of young people would face the trauma Levi had just avoided.

When Hurricane Michael came ashore, it was so fierce that it blew through two states and was still the strongest hurricane to ever hit Georgia by the time it arrived. In Mexico Beach, Florida, about 15 miles south of the base, storm surges threw waves up to 14 feet, which meant the water was high enough to destroy the second story of some of the buildings. The second story of the El Governor Motel, one of the sturdiest buildings around, was swamped by ocean water even while it stood on concrete pilings. Water also wiped out a nearby RV park and carved two new inlets through a state park on a cape farther south, slicing it in thirds.

As the storm thrashed through town, wind stripped trees bare and splintered buildings. Solar panels oddly held shingles in place where roofs were otherwise stripped to plywood. Other buildings were simply flattened. Where some remained aloft, wind and water pushed debris to their feet—fractured boards and parts of what once was—and the still-standing buildings held it there in piles like impromptu barricades. Along the beach itself, the road gave way. And for a while water just stood, flowing in and out of the windows of homes.

Across the whole of Hurricane Michael's path, the force of the wind alone damaged more than fifty thousand structures, three thousand of them for good, swiping roofs and porches off buildings, smashing some to shards, and leaving trees felled in every direction. Wind also derailed 138 railcars. People who were anywhere near the storm described the sound of the wind screaming, the hurricane like an angry train. Sailboats were thrown from a marina into a heap in Panama City, where power lines were tangled and trees covered the streets.

On its inland march, Hurricane Michael dumped massive amounts of rain and spurred several tornadoes. It also flipped a silo, collapsed the roof of a tire store, and knocked down the exterior wall of a building in Marianna, Florida. Power failed across wide swaths of the panhandle, with lights out across the Florida shore and the southern part of Georgia for nearly 200 miles.

Then there were the floods. A state road flooded, overtopping a bridge, and the Chipola River flooded, too, swamping out fishing camps. Once Hurricane Michael was about 160 miles inland, it unloaded 7 inches of rain in a few hours in Quitman County, Georgia. On inland farms, it wiped out peanuts and cotton and, everywhere, again, the trees. The Florida Forest Service and the Georgia Forestry Commission would later estimate that Hurricane Michael downed so many trees that the states' combined timber losses hit $2 billion, and it damaged more than 5 million acres of land. State officials cited another $240 million in replanting costs for trees and fretted over the potential for wildfire while so much kindling lay on forest floors.

The Category 5 winds lasted only three hours in Bay County, a fraction of the forty hours such winds would blow when Dorian hit the Bahamas a year later. Once it stopped, those left reeling would face lives remade by the force of the hurricane. A great many of them would be children.

Aside from the immediate impact of such storms in tropical zones, damage can stack up when these storms are ongoing, and Levi's barrier island is one example of a landscape that is slowly eroding as conditions worsen. Intermittent storms overwhelm the systems that are there to deal with heavy rains. And when that happens, the storms wreak havoc on everything, humans and benthos (the tiny organisms that live at the bottom of the water) alike.

Because the island is small, for example, the potable water is pumped from the mainland. When a storm is headed for the island, the water is turned off to protect the water supply. That's one reason evacuation is standard during storms. And why during one recent hurricane, a house that caught fire in Levi's neighborhood burned to the ground. Sewage flows the other way, in pipes back to the mainland. These pumps don't work so well when the island is saturated or the power is off, so the sewage either backs up into homes or gets dumped into the Indian River Lagoon and circulated back to the Atlantic. The bacteria then mixes with things like lawn fertilizer and leaking septic tanks and makes for algal blooms. This is partly because as temperatures rise, "there's no winter anymore. It doesn't get cold. And because it doesn't get cold, the bacteria doesn't get a chance to die," Leigh-Ann says.

This is obvious enough when Levi and Leigh-Ann and I make a pit stop at a canal along the Indian River Lagoon, where sewage overflows when the city's sanitary system is besieged. It's easy to see the problem just from looking at the water: it's dense and dark, something I've seen before in static canals where sewage outflows pour human waste. Such water also tends to be plagued by the lawn fertilizer people cannot seem to resist on water-fronting acreage. Both the sewage and the fertilizer promote algae growth, which means the water is probably becoming eutrophic while the algae eats up all the oxygen. In other words, it is probably suffocating everything that lives there. Few animals can thrive in water like this. Already there's been a significant die-off of the seagrass that used to grow, up to 100 percent in some areas, eliminating food for manatees, which otherwise like the warm water and shelter of the lagoon. And likely as the algae dies off, it will sink and stack, filling in the canal from the bottom up. A future of dredge projects and pesticides.

There's nothing but a concrete dock to stand on as we overlook the lagoon, so Levi has only a few palm fronds from an overhanging tree to distract himself with. He promptly starts playing Tarzan. As he swings from the ledge of the canal over the dock and the water and back again, there is a minor panic on Leigh-Ann's part about whether he will fall in. This is not because Levi can't swim—Levi is a competitive swimmer and good enough at it that he competes against older kids. The issue is that the water is filthy. And beyond not needing a case of *E. coli*, Levi is wearing Leigh-Ann's shoes. They have the same size feet, they tell me, though hers are wider. So they often trade shoes, since both wear sandals and it often doesn't matter whose are whose. Except now.

"Do not get my shoes wet with this water," Leigh-Ann says, stern, as Levi swings again over the canal and nearly falls.

Levi used to wade in this water as a baby, before it was so routinely polluted. Now he doesn't, though he points out the home of his friend's grandparents across the canal, where he sometimes plays and has fallen in. He has never liked this water, he says later. "It always freaked me out because I couldn't see the bottom."

This lagoon system is important to wildlife, though. Because it is sheltered, "it's a great nursery for young sharks and small fish, and multiple different animals use it because it is a safe haven for them," Levi says. In fact, more than three thousand species of plants and animals live here, with dozens classified as endangered or threatened, like the manatee.

Manatees and juvenile turtles are affected by this pollution, especially when seagrasses are choked out of the water. Sea level rise is already eliminating their habitat, and rising temperatures and ocean acidification strain the species too. Last year, 11 percent of the sea turtles that died or got sick or injured in Florida did so in this region. Brevard County, where Satellite Beach is located, also has the highest number of manatee mortalities in Florida, though the lagoon is a critical habitat for the threatened sea mammal, and the area's beaches are some of most important nesting beaches for sea turtles in the world.

Levi spots a bungee cord with three tennis balls attached to it and drops out of the palm frond to claim it. Next he finds a few rectangles of duct tape, and pulls these apart until sand falls out. As temperatures rise in this region, Leigh-Ann says, the inability of cold weather to kill what bacteria lets loose brings a whole other problem, the much more serious problem of rampant bacterial growth, and Levi especially doesn't enter this water since flesh-eating bacteria were found here.

"If you swim in this with an open wound, you could catch all kinds of who knows what," Leigh-Ann says.

But for an eleven-year-old who's been raised in this ecosystem, life a skinny stretch of land between the beach and a lagoon, canals in between, it doesn't come naturally to Levi to stand aside. There are rowing boats nearby, and he says he plans to try crew when he's old enough. He's only in the sixth grade, and though he's farther along academically, he has to wait until he is twelve to learn. I ask Levi if he's a good swimmer, and he's modest. Says he doesn't know. But Leigh-Ann says when he competed in the rec league, he beat everyone all the time. Now he swims six days a week with the older group, two hours a day. But all that skill, it doesn't help him here.

"Do not fall in that water," she says. "I do not want you touching it."

The devastation Hurricane Michael causes will last for years. Nine months later, Bay County, Florida, home to both Panama City and Mexico Beach, will still be struggling with how to deal with it. In the interim, the horror will be measured in numbers.

$25 billion in property damage.

$4.7 billion in damages to the Air Force base.

31 million cubic yards of debris.

72 million tons of fallen trees.

Twenty-two thousand people displaced.

Five thousand of those made long-term homeless.

Twenty-five deaths.

Two major fires.

And $1.8 billion needed to rebuild the towns. Water mains and towers. Sewage treatment plants. Internet. Roads. Bridges. Stormwater drainage. Community centers. Government buildings. Park, beach, and reef restoration and reforestation too. Even the crayfish need a safe place, their habitat now in the crosshairs of so much rebuilding.

Owing to their youth, the children who lived through the hurricane suffer in compounding ways.

Theirs is exactly the kind of outcome Lise Van Susteren, the trauma expert in the *Juliana* case, warned about. A psychiatrist who co-convened the first conference on the mental health effects of climate change, Van Susteren told the court that flooding and violent storms increase incidents of anxiety, depression, and post-traumatic stress. Those disorders can be especially hard on children, whose brains are still developing and at risk of pumping unnaturally high levels of cortisol, a stress hormone. Their brains can become hardwired this way, producing mental health conditions that can persist throughout life.

That the children who survived Hurricane Michael were still stressed the following summer was obvious. In the Bay County School District, 13 percent of the students were displaced to somewhere else. Of the kids that remained, five thousand of them were still homeless or in temporary housing. And at least 122 children became so mentally unstable that the state took them into protective custody for evaluation for fear they would hurt themselves or someone else. The math as to the exact number of such children was fuzzy. That figure—122—was a statistic assembled by people in a system in crisis. As that crisis slowly resolved, or at least became a new kind of normal, the system morphed into something new, so what numbers were next available arrived by new math, making it years before the number of children seen to be in crisis could be statistically verified as an anomaly. Professionals like Ken Chisholm, the clinical lead for the Bay District Schools mental health team, suspected the number of children in acute distress was higher than normal, higher even than 122. But anecdotally, the impacts were obvious.

"With younger kids, you'll see the triggers start to occur a little quicker. If they lived through the storm and watched their roof come off—which I have had so many kids talk about that experience—anytime you get a thunderstorm, these kids are in a panic," says Chisholm, adding that these effects emerged after the initial shock of trauma was gone, taking root as longer-term anxiety and depression. "You have lots of folks that were living in mobile homes that were knocked off their foundation while people were inside. I had plenty of folks say the window blows out in their home, and so you have glass and you have water pouring in at 150 miles an hour. Just those kinds of stories over and over again."

After the storm, these children were at the mercy of an adult world that was quickly thrown into chaos. Adults were out of work, living in FEMA camps or in collapsing homes or alternate housing, with families stacked on top of each other. There were fewer childcare options, and many of the stressed parents who were unable to take breaks began abusing drugs and alcohol. Hurricane Michael wrecked much of their support—a detox center and a substance abuse center, AA programs that could not run for lack of buildings—and stalled programs for moms with addictions. Forty-seven churches were also damaged or destroyed, leaving the faith community and the support services it offered hobbled too. The lack of housing, doctors, and childcare soon gave way to an uptick of domestic violence, child abuse, drug abuse, and suicides—all things mental health experts have observed in communities affected by extreme weather.

Like the adults, the children had few of the old places to go to process all of this or to just hang around and find a bit of normal. The movie theater was gone and so were the bowling alleys, several parks, and a chunk of the mall. Many of the after-school programs reduced their programming or shuttered completely. The Boys and Girls Club and Girls, Inc., were running at half time and operating out of school buildings, their own facilities destroyed. Some of the sports leagues had also broken down.

This is the kind of fate Levi has always avoided as an evacuee. His home has avoided it, too, through luck and preparation, at least so far. But hurricane season arrives annually in Florida, from June through November. And as more intense hurricanes approach Levi's home from the Atlantic, the likelihood he will experience a devastating storm only increases. Some people never recover their psychological well-being after these kinds of experiences. And each successive experience makes them more vulnerable to the next. When there isn't enough time to recover between storms, that takes its toll too. And because climate change increases incidents of extreme weather, human suffering can only compound, or worse, lead to "pre-traumatic" anxiety—the fear that another storm is just around the corner.

"Children have less life experience than we do. When something like this happens to them, they don't have the ability to process it as well as we do. We have the life experience to know, for example, 'Oh, this is a once-in-a-lifetime storm,'" Chisholm says. "Let's say they go into a place and three hours later they walk out, they don't recognize anything. That's traumatic. And then you have those kids who thought they might die."

All of this made the shift back to normal rough on the children who survived Hurricane Michael. And normal was a new normal. The old life was gone. In addition to the shifting routines at home, the shifting locations of home, some elementary-aged kids were hit with school closures, too, so that many of the community's youngest children were displaced both at school and at home. These students had new teachers, new classmates—a whole new social order to learn. Plus the teachers were as stressed as all the other adults. Many of them were displaced, too, and experiencing a kind of malaise about supporting all the other people who needed support while they faced their own impending relocations and rising rents as housing tightened. The official term for this is "compassion fatigue." It is a tough condition for teachers whose salaries hover at $30,000 a year, a figure even a job at McDonald's can rival.

The adult world bearing down, the kids' own lives disrupted, many young people were turning up at school only to act out or have their minds wander. The adolescent mental health treatment center was among the places wrecked by the storm and could not help them.

No one could say for sure what would happen to these children in the long run—only time and life could really tell. But studies of youth from Katrina, from Hurricane Charley, from Superstorm Sandy, all show disaster trauma can have lasting psychological impacts for kids. There is a bell curve, it seems, of possibilities. On one end, Chisholm notes, are the youth who acclimate—like our ancestors did to frequent deaths of their children, to disease, to lack of medical care. On the other, there are those who fail to cope, end up psychotic or suicidal. And then there are all the outcomes in between.

Asked whether the *Juliana* plaintiffs were being harmed, and interviewing each, Van Susteren called the group of youth, with their varying experiences of climate trauma, the "Climate Cassandras," a call back to the Greek myth of Cassandra, whose gift was to see the future but whose curse was to not be believed. Van Susteren theorized the *Juliana* youth and climate activists like them could grow up like the children of alcoholics, constantly repeating their alarm without result, reaction. They would learn to be grown up before they had a chance to be young. Their lives would be attended by the guilt and frustration of not being able to stop climate change. And meanwhile they would shoulder the "debilitating knowledge" that humans are the cause of it. Existential questions about species survival and mass extinction would walk through life with them. And they would bear the sadness of knowing "that we did not value them enough to bother protecting them from harm," even though they warned us.

Levi will join the youth leadership program at Toastmasters soon after. Go on a speaking tour of Unitarian Universalist churches in the Pacific Northwest, knock out speech after event in Florida, keep doing his media interviews. Still, he'll face another hurricane within a year. After Dorian makes for Florida and the president adjusts its course with a Sharpie, asks whether we can just nuke these storms and be done with them, it will wipe out chunks of dune above Levi's Pelican Beach, cover the beach with seaweed, and carve sea turtle nests in half along the sand like a real-life diorama. Afterward, we will talk on the phone and he will tell me he was amazed and depressed by this all at once. But on the day of Hurricane Michael's landfall, Levi hasn't seen any of this devastation yet.

In these days when he is still eleven, on our visit in October 2018, we end up at a spot where salt water flows off the Atlantic to make the Indian River and the lagoon system still one of the most diverse in the Northern Hemisphere. Levi pops out of the car wearing the tennis balls and bungee cord garland he salvaged from the canal. In the time it has taken to drive from here—five minutes, probably—he has wrapped the bungee portion around himself so that the tennis balls ride along the front of him like a very large necklace. Leigh-Ann is aghast when she turns from the driver's seat and sees this. There is a brief debate about the probable bacterial content of this ornament. And then Levi is forced to undecorate himself.

We're on the south end of the lagoon, where the wind is fierce, the fetch incredible, and the tide rises up. For the most part it is ocean, an inlet that peels off the Atlantic and wraps itself around the island's western side in a series of sheltered lakes and canals. Levi walks to the end of one of three wooden piers and watches as a couple tries to launch a skiff in the rough water. There's no pushing this boat against this surf, not from the shore, so the duo take it to the end of the dock, pulling it by a rope, and try to climb aboard from there. It works, but only on the second try, and Levi stands back as they struggle onboard and motor off, next finding a dead catfish lying at the foot of the surf. He hoists the fish on the end of a stick, looks out at the water, pondering.

"I think if you throw him in, he's going to come back to you," I say.

"I'm going to try that," Levi says, and runs to the westernmost pier to fling the fish into the sea.

"We had a fish kill a couple years ago where there was a million dead fish here. Just everywhere, dead fish," Leigh-Ann says. "Some manatees were dead. Some dolphins. Some other bigger things, birds. It was horrible." She says the smell was terrible, too, and that all this was caused by a combination of hot weather, a storm that washed nutrients into the water, and the resulting algal bloom. Climate change just makes all of these things worse, she says, exacerbating problems that are already here.

For example, a boat is sunk about 20 yards off the pier even now. The bow is slightly nosing out of the water as if gasping for air, but the stern is done for, submerged, sunk. It's a sport boat, and the water is rising and falling over its portside windows. Sometimes in storms, this happens, Leigh-Ann says. Not infrequently. And when boats sink, the fuel lets loose and becomes just another pollutant in the water. So, too, does everything on board. I imagine coolers of beer. Towels. Plates and silverware. A radio. This is not a huge vessel. But it's big enough for a galley and the amenities pleasure boats often have. Refrigerators. Microwaves. Stovetop ranges.

A driver stops to ask if the boat just sank or has been there. Then a fellow enviro stops by to tell us the Coast Guard was able to get the fuel off the boat before it spilled into the water. This and details of a recent vote of the county commission to ban septic tanks next to the lagoon. Leigh-Ann invites the man to Levi's upcoming send-off at the church. We don't yet know that the trial will be canceled. That Levi won't be living in Oregon this fall in the care of other parents, grandparents, and the staff of Our Children's Trust in the stretches when his mother can't be there.

The dead catfish washes back ashore. Levi names him Leroy, then shows me a few more sea beans and a huge snail shell. We see a crane on the pier then, and soon Levi seems to be having a kind of conversation with this bird. I remember how much this is possible for children, and when he leaves the dock for the parking lot and the crane follows, Levi reminds me how much closer children can be to nature than adults. How much they can feel the life of an inchworm, of a beetle, a bird. While Levi stands near to his mother and me, starts to imitate the body posture of the crane, first he is just turning his body toward it, somewhat subconsciously, as the crane walks around us. But then the crane takes on a funny pose, as if bracing for wind, a leg raised, and Levi starts to do the same, trying on this one-legged stance. You can hear the wonder: What is this bird doing? Levi's head is tilted. And because Levi is small, the crane nearly as tall as him, next it is the crane's turn to be curious: What is this yellow thing with the blowing hair? Levi moves away, back to the pier, and the bird follows. Soon they are on the pier, one pacing one way, the other following, and then the reverse. It is like this for a while. A slow back-and-forth between two same-sized creatures who don't know much about each other but seem to want to learn. Or at least play.

Leigh-Ann gets a call: press staff from Our Children's Trust. Can Levi do an interview on NPR? As the logistics unfold, and Levi and the crane do their thing, the bird gets too close to me and flies off to the next pier. Levi walks back toward the water, spots a dolphin and points it out.

"I've seen tons of manatees," he says. "I think they're beautiful." He says they're slow and lazy, "but when they want to go, they can really go." In floods, he has seen the manatees crawl onto the banks to eat people's lawns to avoid starving, their heads straining for the grass.

After I leave, dodging Hurricane Michael airport by airport and learning, for the first time, about the dozens of people killed, I hear Levi talk to Terri Gross, flawlessly, on NPR. Despite already having met him and knowing how eloquent he can be, I am baffled by his poise. It happens again when I speak to him for Reuters in front of the courthouse after trial is canceled, this bafflement. Though he is only eleven, Levi's expertise at speaking in short, quotable sentences that exactly respond to the questions he is asked is well honed. I realize that because he is the youngest plaintiff, he is among the most frequently quoted, certainly the most photographed, and is rapidly becoming a highly skilled public speaker. He is growing up fast. And seeing this alongside the horrors that threaten him makes me wonder what climate change will take from him first: his innocence or his landscape.

CHAPTER EIGHT

"The Other Side of the So-Called 'End of History'"

People say to Levi what they say to all the plaintiffs: "You're our last hope." But Levi is eleven. He is only half as old as Kelsey, the oldest plaintiff. And Kelsey is only half as old as me. By the time we reach the IPCC deadline to take action to halt climate change or bust—2030, when unchecked global warming will usher in an era of food shortages, worsening wildfires, mass die-offs of coral reefs, persistent droughts, and rising seas swallowing coasts—Levi will be only twenty-three, barely old enough to drink a legal beer. If he goes to college, he might not have graduated. He may not have a job or housing of his own yet, if ever. What is it that grown-ups think he can do for them? And why haven't they done it for him? Why are adults so eager to foist this baggage on the young, forfeit action of their own?

I spend a lot of time on these questions. I try not to let them come out of my mouth because, according to Kelsey, this type of guilty navel-gazing is insufferable. But it's hard to be an adult in this situation, at least one of the older adults, without feeling like you have your fingerprints on the failed climate fight. As insufferable as it apparently makes me, I want to know how we got here.

I put this question to Julia Olson, a fellow Gen Xer and the lead attorney in the case, after the trial is canceled: Why weren't we in the streets? How did we leave them with this? She says what I remember: "It felt like the threat, when I was growing up, was the Soviet Union, and it was the Cold War, and it was nuclear bombs. That was what my perception of what the threat to the world was until after college."

We are in her office, in what must have been a bedroom once, in a converted bungalow in downtown Eugene, a window looking onto a slim patch of lawn, wood paneling on the walls.

I remember this too. I didn't grow up with nuclear fallout tunnels in my high school, like Olson did, her home in Colorado Springs somewhat neighbors with the North American Aerospace Defense Command or NORAD, the place where satellites and radar scanned the skies for incoming projectiles. But I do remember being terrified of nuclear war, that when we were the age the plaintiffs are now, we lived in fear of some unhinged adult pressing The Button and having to dive under our desks before the shockwaves came.

This day in her office, Olson and I talk about what we knew. How we were aware of global warming but had difficulty parsing it from the problem of the hole in the ozone, a problem targeted by an international agreement—the Montreal Protocol—and then solved. As we got older, it seemed like the problem of the atmosphere stayed solved. When Olson learned about climate change in college, it wasn't taught to her in urgent terms. She became mindful of population growth, became a vegetarian. And that earth-loving ethos carried forward in the form of conscientious consumption, like it did for many people our age. I did what most of my peers did with this: bought recycled toilet paper and reusable coffee mugs and started up with theoretically responsible terms like "replacement population." I stopped buying things that made the world worse—beef plumped full of hormones, vegetables doused in pesticides—and started fending for my health. But while we got older still, past old enough to drink that legal beer and have housing of our own, it became clear that climate change was a distinct threat, and an important one. What wasn't clear was that we weren't doing enough to stop it. That the projections for how much time we had to fix it were all wrong. And that climate change couldn't be solved with the personal choices we were dutifully making at the cash register and by buying the right kind of car.

Now, Kelsey says that when she speaks in public there is often a kind of challenge that comes. It takes the form of a question, but it doesn't seek an answer. It's more like a rant in disguise. A castigation. It is almost always asked by a man, she says, usually a boomer. And whatever the phrasing, the real message is usually one of two things, or sometimes both. First: your hope is your naïveté. And second: if this problem could be fixed, we would have fixed it.

Lately she's been on a bit of a conference circuit: schools and universities, topical events around food rights and fashion. In these places, sure as she moves her mouth, she is often scolded for it. And she hates this. This work is her heart and she wants to talk about it as much as she can, wants to bring people along with her. But she doesn't think it's her job to justify the youth climate movement to older people or to keep explaining why it exists. Unless they are in charge of things or want to stand with her, it hardly matters what they think. They are bystanders now. Most dead and gone by the time the mess they made starts to be her generation's—Gen Z's—to clean. If they want to know, want to understand, she thinks they ought to try to educate themselves. She is tired of doing it.

She brings this up one day over lunch. Looks sideways to a corner of her brain where these thoughts have been stewing and tries to articulate what it is that's been nagging at her. The analogy she uses is this one: it shouldn't be up to people of color to explain racism, or to women to explain what it means to be sexually harassed. And it shouldn't be up to the young to explain their call for a livable planet to the people who are on course to leave it in shambles. She says: don't let your guilt be the center of things when really it is not about you.

Of course, she is right. Isn't this the root of the blowback question, after all? Guilt? And ego? Both take flight as skepticism, as denial, as if the young are ungrateful for the America that is theirs. *Why isn't it good enough for you, young lady, when it was good enough for us? We were hot too. We bought air conditioners.*

Oh, but I see it. The white flag rising. The roll over. The turn away. This surrender, shrouded in its alternating turns of guilt and of denial.

More questions for Kelsey. At yet another podium in yet another auditorium, weeks later, here it comes, the reproach. From the handwritten cards collected from the audience. It happens all the time now. From the microphone, between the extemporaneous soliloquies about how they fought too. Sometimes it happens in the grocery store, in press interviews, on the Internet. People ask her how she feels about a newly conservative Supreme Court. If she really thinks she can win. If her optimism is all that deserved. But sometimes to be successful, she says, you have to accept that success is not a guarantee. Maybe the most important fights don't come with a roadmap to winning. Isn't this the core of her case against the government? That there is no path for redress? Civil rights leaders didn't have a roadmap either. They used the courts because we have a Constitution. It is supposed to be enough.

Kelsey joined this movement as a young person because she was worried about the future. But now that the future is here, she says she sees younger people like Levi join the movement because they are worried about themselves. Already the metric of urgency is recalibrating, decade by decade, even within Gen Z.

Also recalibrating? The guilt. Our guilt is for them. But for Kelsey? "My guilt is a milk latte to go," she says. Every paper cup, every plastic lid a tiny torture. She knows we need systemic change. Still, her part in being a person hurts. So does this judgment about the ways that she lives with it. Until we understand that, she says, we don't understand what it's like to be young in this world, to only ever live amid the threat of ecological collapse. To be young enough to have to care how it turns out.

Those of us who are older, it's not our fault that our government did not say: we need massive, systemic change, and we have to do this now—or else. Or that our government chose to downplay this inevitability. Those facts are betrayals of all of us, not just of them. But when we don't listen to the affected generation now, when we don't acknowledge this neglect of oversight—maybe that is our betrayal of them.

All of us now, in America, face a personal reckoning with the compounding environmental debt wrought by this failure. For Olson, this realization landed hard in a movie theater in 2006 when she was watching Al Gore's *An Inconvenient Truth*. She was pregnant with her second child, trying to escape the heat, and, watching the floods and fires and the other perils of climate breakdown foretold on screen, heartsick.

The reaction she would ultimately have—a strategy to seed fifty-one legal actions in a single year—wouldn't come for a while. First, she had to have that baby and be her best self for that child and the one that came before it, not yet three years old. For her, it was a time of grinding her own grains and making bread and fresh yogurt and doing her best to care for the two young people who were depending on her for love and shelter and sustenance. On the environmental front, she resolved that she would fly less, purge plastic, be a member of her community, and do what she could from home.

But when her youngest was three and a half, successfully weaned, the age of making bread and yogurt began to give way to whatever else could be made from scratch. Now that no one was physically attached to her, didn't physically always need her, she was thinking about what legal action had been taken on climate change, how inadequate it so far seemed. She envisioned something bigger, more coordinated and global, litigation to spur systemic change, not just react to the latest fad in fossil fuels.

This drive she felt, it was both past and prologue. Olson knew what it was like to struggle as a young person, didn't wish it on anyone. She was seventeen, sitting in a car in her neighborhood in Colorado Springs, when a mentally ill man with access to guns shot her in the chest one day in 1988, collapsing her lung and sending her hurtling toward emergency surgery and a new life. What followed then was a week in the hospital, a period of disorientation and healing. Eventually, she rose up another person. And in between there was an infuriating letter from the man who shot her—all defense, short on empathy—handed to her in the moments while she waited outside a courtroom, before she would walk in and tell a judge what she wanted the court to do with this man. In between, no one knew how to help her with how this experience felt.

"Nobody understood what that would mean for me, as a seventeen-year-old, going through that. And *nobody* knew how to help me around that, to integrate that experience and heal from that experience." We are still in her office, sitting now with knees not far apart in a corner of the room where a couch and a chair are squeezed in the space around her desk.

It was a formative time for many reasons, she says. It was the first time she saw the clout that lawyers and judges could have in a courtroom. Felt the power of being asked how she wanted a bad situation to end. Learned who she would be in the world when she told the judge that what she wanted was for the man who shot her to never own a gun again and to get the help he needed so he could come back to the community without hurting anyone else. Henceforth, hers would be a path of compassionate justice. College. Law school. She grew up fast, had always been growing up fast. She learned to fight, to be independent, to persevere through challenges. Now her favorite thing is climbing mountains, standing on top. Along the way, she learned how empowering it is to speak for yourself, and for the world to make room for you to simply say what you need to.

"I think that is one of the things that's so important for young people—whether it is climate trauma or gun trauma or violence, it's allowing them to have a platform and a voice to share their experience and their story and try to educate others and to get other people to try to stand with them. I think a lot about that. I didn't have that," she says.

She didn't know how to give that platform to young people in the climate fight. But she wanted to at least stand with them. So she attended an environmental law conference at the University of Oregon, where she started to get clear on what her vision for litigation looked like. She heard a speech by Mary Christina Wood, the head of the university's Environmental and Natural Resources Law Center and a scholar who had already devised a strategy.

Wood had the same effect on Olson that she would have on Alex Loznak—an intoxicating one. A woman of unconventional thought with a righteous environmental bent, Wood was a child growing up in the sixties and seventies when she witnessed the plundering of the Columbia River watershed. This after the bridge between Portland and Vancouver was built, paving the way for thousands of acres of farmland to become "wall-to-wall suburbia." She would later write in her 351-page treatise, *Nature's Trust,* about how developers tore up the farms and "bulldozers operated from dawn to dusk, demolishing wetlands, creeks, forests, springs ..." Trees, soil, riparian land were collateral damage in the remaking of her world. "McMansions sprouted everywhere, as far as the eye could see, separated only by strip malls pimpled with huge box stores and fast food," she wrote.

Wood declared this era one of legalized plunder. The laws that were supposed to stop it, she later theorized, were a significant part of the problem. Consider Earth Day 1970, when twenty million people took to the streets to protest everything from water pollution to smog and clear-cutting, pressed government to hold industry accountable. What followed was the Clean Water Act, the Clean Air Act, the Endangered Species Act—eighteen such statutes that proponents pronounced a win. But if they could allow the pillaging of entire ecosystems by developers, did they work? Wood did not think so.

"They are the permission slips for unmitigated disaster, as far as I can tell," she says one day, in a conference room at the law center. It's on the second floor of the law school at the University of Oregon, a place with an arched entrance, flat carpeting, and a lounge lit by two-story windows for the students who never seem to stop studying there.

By then, Wood had chalked modern environmental law up to "a failed legal experiment." These laws she refers to, they're called statutory environmental laws. And it is hard to argue with her when she says that instead of phasing out pollution, as intended, these laws opened the door for governments to permit select abuses of nature, a door Americans are still trying to close. In between, Wood notes, the system became byzantine, with one federal manual topping two thousand pages just to explain it. She calls such tomes the handbooks of lobbyists. And says permits are not unique, instead sanctioning assaults on the environment with a kind of humdrum daily rigor. Chemical plants and clear-cuts. Mountaintop removal mining and the spreading of chemical waste on farms. Sewage and toxins dumped into waterways. All while society is numb to these things, convinced the law is working.

"Statutory law isn't the only law," Wood tells me. "There's another premise, or a more foundational set of laws or body of law called the public trust. And it's preconstitutional. It's got constitutional force and it actually should channel the discretion of the executive agencies. They can't just do what they want."

What she's referencing is called the public trust doctrine. And in her scholarly work, Wood has excavated this doctrine from all the rest. That day in the law school, she likens the work to pulling the ivy off the useless other law that covered it. She says its roots date to 535 AD, to the Roman Institutes of Justinian and the Magna Carta of England, to the courts of the United States and beyond. Once, it was a thing that made democracy distinct from oligarchy: to have a government that protected the assets of its people. Some legal thinkers, Wood among them, say this public trust ideology is what America's founders carried to the New World in their boats. They used words like *posterity*, which implied they meant for things to be around awhile, sustaining one generation after the next. Thomas Jefferson, for example, described in letters to John Taylor and James Madison how when it comes to nature, each generation has equal right "to the free possessions of the earth . . . for their subsistence, unencumbered by their predecessors who, like them, were but tenants for life." Or, in other words, as Wood put it, "one generation could not bind another with natural debt."

She assembled all of this—the letters, the case law. And she found that the American courts had tested some of it. Enough to decide that the federal government ought to maintain the waters of the nation in some sort of navigable state, and that the state of Illinois should not be in the business of handing over the shores of Lake Michigan to a railroad company. And even though the nation has since been mired in statutory law, handing the authority to protect the environment over to the states, who have handed it over to bureaucrats, Wood says reasserting government's role to protect the wilds is doable. Anywhere circumstances have crossed boundaries, say a river, or where the scale of threat has spanned states, or disputes have arisen over the shared resources of nations, she says the case has been made. Oregon used a similar rationale when its Supreme Court made the state's beaches public. So have other nations relying on international law, like the Supreme Court of the Philippines when it halted logging for fear of the forests vanishing, and the Supreme Court of India when it asserted trust over natural gas reserves to maintain a supply for the Indian people. Just because the responsibility to protect the environment has been abdicated by Congress, by courts, by a disengaged public, and by executive agencies that have more or less fallen captive to industry, she says, doesn't mean the law can't still force it back into place. From her perspective, this means using the public trust to accommodate the things we now know people need to survive: biodiversity, groundwater, forests, healthy soils. And, yes, a stable climate.

To her detractors—and there are plenty—Wood wrote in her book, "No law should be defined by its violation," and suggested maybe the biggest stumbling blocks to change are activists themselves. She thinks all the people who are calling for change, and have been doing so for years, are "so demoralized by the overwhelming corporate influence on government that they find any transformative approach unimaginable." She wants to unstick them. Unstick all of it.

Hearing these ideas, Julia Olson was hooked. That Wood was a mother in the same stage of mothering, her and Olson's children all of an age, made for a certain synergy. Within a few weeks of meeting Wood, Olson was fomenting all of this into action. Her birthday was in April, so she did the thing she always did: took a hike with a group of friends, climbed Spencer Butte south of Eugene and hung out on the rocky summit overlooking the valley until it was dark. At the top of the butte—as she liked to do each year—she set her intentions for the next. That's when she said it out loud.

"I said, 'This is what I'm going to do. I'm going to do this big whole thing and we're going to file cases everywhere around the world,'" Olson says later.

She intended for it to be a pro bono endeavor. A volunteer mission. But it turned out that filing dozens of lawsuits requires at least a small staff. Which requires at least a small budget. Which requires some sort of organization. So Olson founded a nonprofit. She called it Our Children's Trust, a term culled from Wood's theories, to underscore the notion that children have a vested interest in nature, and that it's up to the government to protect it for them. Olson and a core group of female colleagues joined the inaugural board, made it official.

This is how it came to be that this movement of youth lawsuits was founded not by funders or by activists or exclusively by youth. And not—as conspiracy theorists would later charge—liberal ideologues bent on infecting the youth with talk of the climate hoax. But by a handful of super-pissed-off moms and aunties who also happen to be lawyers. That several of these women are Gen X moms whose Gen Z kids will grow up to face the consequences of the climate fight—won or lost—is no coincidence.

Filing all the lawsuits took another ten months. Lots of attorneys, all of them pro bono, with Olson minding the deadlines and the consistency of the cases. A series of strategic meetings came next. The first brought roughly thirty attorneys, about a third of them from abroad, to a bland conference room at the Phoenix Inn hotel in Eugene. There they considered whether to sign on as volunteers, sue state and federal governments on behalf of the young. It was mostly Olson's job to convince them, so she invited two persuasive speakers. One was James Hansen, the renowned former NASA scientist who had testified to Congress in the eighties that climate change is real, man-made, and urgent. The other was Alec Loorz, age sixteen.

Alec, with his mother, was the founder of Kids vs. Global Warming and iMatter. He was not a heavyweight by James Hansen's standards, but like Xiuhtezcatl he was an early youth advocate in the climate fight, known to a lot of other kids. Plus, he could talk. At sixteen, Alec's voice was already a commanding one, and he had the poise and conviction of someone who spoke from the heart. His ability to understand what was happening to the environment and to translate it into a message that inspired people was especially unique, both for the person and the era. He was so effective at this that in the six years he was active as a youth advocate, Alec gave more than five hundred talks, had to leave school in the eighth grade just to accommodate all the requests.

He was a prepper, so he had a PowerPoint. And after Hansen explained the science to the roomful of attentive attorneys, Alec walked to the front.

He doesn't remember the exact words that tumbled out. But he remembers well what was on his mind then. How he was learning, like a lot of other youth, that he was not simply a powerless juvenile with ideas that no one wanted to hear. But rather, precisely because of his age, he was able to claim a moral authority to speak about the future because he was among the young people who would inherit it. As a public speaker, he sometimes felt short on tools—a voice was only that. But he also knew his message was a powerful one. And what was obvious to others was that he was born a person who could wield words like weapons, send tornadoes through people's minds. When he heard about the lawsuits, they struck him as a way to amplify what it was he and other young activists were trying to say.

"It could have been that I was the one . . . who brought that point in a clear way about our future, that we as youth have a right to a livable planet, and this is a way to actually make our voices heard on a scale that's larger than we're able to otherwise," he tells me one day on the phone, reflecting back on this story now nearly a decade old. "I can definitely remember that feeling, the sense of, 'Oh sweet, you guys can help. I've been trying to do this as a public speaker but if you're able to do this, if you all are lawyers and you actually have an approach, I know young people who would be plaintiffs.'"

What the lawyers heard was the momentum of his voice, and a request as clear as anything: We need you. Please help.

"All these attorneys sitting there and hearing this young person . . . just blew everybody away," Olson said.

Every attorney in the room signed on. And Alec became the first plaintiff in a federal lawsuit that challenged government *inaction* on climate change. He also used his extensive network to help recruit many of the kids that became plaintiffs in state actions, including Xiuhtezcatl. Not all were lawsuits—some were state-level rulemaking efforts, but legal matters all the same. Around Mother's Day 2011, all of them dropped at once. Fifty-one in America, targeting all fifty states and the federal government.

Alec's case didn't last long by legal standards. Two years. And Alec left the movement soon after, grateful for the experience but burned out by seventeen, exhausted by the effort and by the weight of such a heavy psychic burden. He emerged years later, an aspiring artist and filmmaker intent on observing the subtle changes of landscape. With his camera he could illustrate, it seemed, the connection a person could have with a place. He believed this was the piece the movement had missed: that at our core, humans are nature and nature is us. As with Levi and the crane, Alec sensed a conversation brewing that would never be had in words. He turned his focus to expression, to art, and found a new peace in being able to express himself with something other than that incredible voice.

The legal loss was a setback for the attorneys. "The feedback we got back from the courts was you need to challenge something the government is actually *doing*, not its failure to act," Olson says.

Undaunted, they integrated that feedback into new strategies, new cases. And started again.

———

The next lawsuit would be brought on behalf of the *Juliana* twenty-one, a group assembled largely by Kelsey and Xiuhtezcatl, who were already plaintiffs in state lawsuits in Oregon and Colorado. The group also included several youths who'd reached out to Our Children's Trust on their own or participated in its camps.

Five years earlier, Olson had recruited Phil Gregory to the legal team. A trial lawyer with a history of crafting battle-ready plans for big corporate lawsuits, Gregory organized the attorneys according to the same playbook the government later appeared to have used—the Big Tobacco playbook. He knew tobacco litigation was successful for one reason: because somebody leaked the documents showing that executives knew smoking caused cancer and was addictive. He wanted to do the same thing for the *Juliana* case: prove the government knew about climate change and knew it posed a threat to its citizenry.

Every plaintiff made their own choice about what roles they wanted to play—talk to the press and the crowds, stand down on the extracurriculars, or roll up their sleeves on research. Alex Loznak, the southern Oregonian whose family farms hazelnuts, would be unique among the plaintiffs in his passion for the legal work. His first year at Columbia behind him, Alex was increasingly drawn to the law, and what he wanted was an internship. To dig deeply into the legal aspects of the case and garner trial-level experience as an undergrad. So he got it.

Along with the lawyers, Alex was among those dispatched on a hunt for evidence. His particular charge was a search of presidential libraries. He scoured letters. Memos. Reports. Index by index. Box by box. Science. Meteorology. His task was to ask two questions: What did the government know? And when did the government know it? He combed the Eisenhower library in Abilene, Kansas, an Old West kind of town where volunteer actors reenact cowboy gunfights. Then the Truman library in Independence, Missouri, where Truman trained his own docents and sometimes liked to answer the phones. Then the Kennedy library in Boston, where he found *the* letter.

"It's the first known White House mention of climate change," Alex says, grinning later at the kitchen table in his childhood home. The letter is from 1961. He calls it "my little footnote in the history book."

It's an exceptional footnote. The letter is a response from President Kennedy's assistant to New Mexico senator Clinton Anderson. Anderson had written to Kennedy the week before, urging him to read an article published in *Fortune* in 1955, in which a brilliant physicist and mathematician who had worked on the Manhattan Project used computers to model the effects of a warming globe on glaciation, precipitation, sea level rise, hurricanes, and the habitability of the coasts.

"It's a really broad, sweeping article that talks about just various impacts of technology on human society and it sort of predicts all of these doomsday scenarios," Alex says. "He talked about automation and how automation was going to eliminate jobs. This is in 1955 and now we're talking about it, so he had some very good predictions. But one of the things that he predicts is carbon dioxide emissions are going to cause warming and it's going to cause Greenland and Antarctica to melt and cause 15 feet of sea level rise. This is in 1955. And he also said that the earth had already warmed 1 degree Fahrenheit at that point in 1955 as a result of carbon dioxide emissions."

This was at a time when policy makers were fleetingly absorbed in the prospect of making rain, the question of whether they could manipulate weather for farming and other purposes. The article posited maybe so, but cautioned that weather was a product of solar energy absorbed by Earth and subject to delicate influences. That to mess with it would affect the whole world, merge interests between nations more than the economy or wars ever had.

Kennedy's assistant responded with thanks for the letter and article. When Alex found both, he was inside the library. Built on what used to be a garbage dump, it stood on Columbia Point in Boston's Dorchester neighborhood, overlooking Dorchester Bay through a 115-foot tower of gray glass. As Alex read these 61-year-old words, words warning that a melting Greenland and Antarctica could cause massive sea level rise, he paused to watch the water lilt not far from where he stood. It provoked "this eerie feeling, looking out across the waters and seeing, well, you know, that may come true."

This typewritten exchange between Kennedy and Anderson is now enshrined on a timeline of what-the-government-knew-when that rings the walls of Olson's office. It is handwritten on a massive strip of butcher paper, brainstormed over with Magic Marker, a mural of evidence that starts just to the left of the door and traverses first a plaster wall, then the wood paneling of the next, wrapping around Olson's desk. On the whole of this banner, probably 25 feet of it, this letter that Alex found represents a single bullet point. One. Tiny. Bullet. Point.

The rest are mind blowing. Svante Arrhenius's 1890 discovery of the greenhouse effect. His clear warning that the world's industrial processes emit carbon dioxide that causes global warming. Confirmation of that theory from Guy Callendar in 1938, when Roosevelt was president, and Callendar's assertion that humans were an agent of change. The first days of government-sponsored carbon monitoring in Hawaii in 1958. The subsequent scientific focus on what all this carbon dioxide would do. The worry that it could cause more than 3 degrees Celsius of warming. The discovery by 1969 that doubling the carbon that had already accumulated would cause 2 degrees Celsius of warming. The understanding that that was bad. A 1965 White House report in the Johnson administration with a whole chapter on atmospheric carbon. A National Academy of Sciences report projecting a 25-percent increase in carbon accumulation by 2000 without change. Nixon's national Energy Future report, recommending a national shift away from fossil fuels and toward renewables in 1973. Ford's State of the Union address in 1975 calling for reduction of oil imports, development of alternative energy, and establishment of pollution standards for cars. The start of the ethanol fuels program that followed. Efforts to contain pollution. A 1977 report from the Government Accounting Office that life on Earth would be harmed by 2 degrees Celsius of warming, urging limits on atmospheric carbon over the next one hundred to two hundred years. The 1983 EPA report recommending an end to coal burning by 2000, a ban on oil shale, a 300-percent carbon tax to halt global warming before it reached 2 degrees Celsius by 2040, and up to 11 feet of sea level rise by 2100. Then the retrench. Reagan.

A new focus on fossil fuels. A Department of Energy research program on fracking. The ensuing march of more of the same—Bush, Clinton, Bush—until Obama made the United States the number one producer of oil and gas in the world, bragged about it, even as he warned of a changing climate.

The first time I see this, see government knowledge of climate science and understand the depth of the science itself, absorb that the list is so long it wraps around a room, I am thoroughly, incomprehensibly angry. This is not so typical. Despite all charges to the contrary, twenty years in journalism makes it very hard for a person to get worked up. Every day is human foible, and much of it looks like the same kind of foible, over and over again. This, though, this is different. When I see it, I feel hoodwinked, bamboozled. Like I spent years watching the wrong hand of politics while the other was doing magic tricks.

It strikes me then why so many people think climate change is a conspiracy. Because when you have a government in possession of such extensive scientific knowledge, and that government is abdicating responsibility to act while its leaders pander to moneyed special interests, and those interests get ever more moneyed by seeding doubt about the need to ever do anything, that *actually is a conspiracy*. And it is not a cabal of liberals in cahoots with academics in cahoots with the media, as the worst of conspiracy theorists will tell you, thanks to all that money. It is much simpler than that. It is just rich people helping other rich people get richer by forestalling action on a difficult problem that the government really ought to be solving. The Great Recession should be evidence enough that the American establishment will push its citizenry to the brink of ruin. And also that Americans are quite capable of whistling their way straight off a cliff as long as they are following the money.

I hate to admit my ignorance. I really do, because it makes me feel like a fool. I knew that oil drives wars and elections, drives the politics of the Middle East. I knew that climate change is the result of fossil fuel consumption and carbon dioxide emission. That it is real. That we need to do something about it. But I didn't know that my government saw climate breakdown as imminent, did not heed dire warnings for decades, meanwhile supported policies and industries that worsened this problem, and also engaged the nation in a debate about whether the problem even exists, eating the clock while it counted down to a crisis that only compounded.

In soccer, this period of endgame dawdling is called garbage time. When I look at these walls, see this enormous banner, I realize I have been living my entire adult, voting life in garbage time.

Until this moment in Olson's office with the timeline, my sense of why Gen X had lost focus on the climate fight had more to do with resignation. I thought that novelist Douglas Coupland had been right about Gen X all along. That we arrived at adulthood thinking we'd already failed and just stayed that way because nothing we were taught to expect about adult life in America was really the way it turned out to be. No laugh track, few self-fulfilling careers. Little financial security, and worse, midcareer when the economy upended, sandwiched between the parents and the kids with a wildcard case of layoffs and furloughs.

Now middle aged, Xers are a generation of cynics. People who struggled to find purpose in the vast landscape of shopping malls and convenience stores of their youth. Caught somewhere between knowing their good fortune relative to the young and feeling screwed all the same, screwed by this culture that commoditized everything or threw it away, by our parents' generation that held onto power until Xers were simply skipped over; Gen X is a generation that would never execute its vision for leadership so never had one. Maybe Gen X should have inherited more of this nation, assumed the environmental mantle. But what we have instead are halfhearted stabs at Earth Day, recycled toilet paper, the locavore movement, and the mild commercial success of the electric vehicle. These are moderate gains, slight improvements on what came before. But in many ways they are like the clapper, like reality television—relics of a generation raised on suburbs and detention, on sugary cereal, to questionable effect.

Somewhere in between, the Internet came along and X made the world a little smaller—created YouTube and Amazon, Wikipedia and Netflix, Google—as a way out of America's suffocating cotton-candy culture. But as writer Jeff Gordinier would observe, Gen X figured out how to connect but not how to join, too antiestablishment. Reflecting on Henry James's novella *The Beast in the Jungle*, he wrote, "The Beast was inaction. . . . The Beast was *the very act of sitting around waiting for something to happen*." And the inaction was us.

We never did what these kids are doing. We didn't take to the courts to demand something better. We didn't run for office—not often—and win the grassroots vote. The few Gen Xers who felt the call to urgency, who stormed into action as the Earth Liberation Front, burning ski resorts and research facilities, torching SUVs, they were hunted by the second Bush administration and made examples of, sentenced to federal prison for far longer than a lot of people would have believed. Calling them terrorists in a post-9/11 world was enough to erase what they had to say. To make them enemies of the state and their tactics a thing no one else would try. Maybe the rest of the Xers fell into line with this, realized that resistance was futile, and did what we were trained to do all along: watch, disaffected, change the channel, recede.

Maybe I am insufferable, but all of this bothers me. Don't believe me? Consider this: 75 percent of people eighteen to thirty-four were worried about global warming in a spring 2018 Gallup poll. That number for Gen Xers was 62 percent. Boomers? 56 percent.

Historian Bill Kovarik puts it into perspective, though. He reminds me that what Olson and I remember from our youths—the panic about nuclear war, the sense of horror at the adults in charge—was society recalibrating the wrong turn the climate fight took, the one we're not old enough to remember. That climate change was indeed clear to past presidents, and society too. That some effort was made at solving the problem. That the security community drove some of those efforts, aware of the political instability climate change could cause; and that even ordinary Americans got a whiff of this with the oil embargos of the seventies. Some thought the answer was nuclear power, all while renewables were more palatable and not such a bad idea either. There was talk about a country that could be more self-reliant, more energy independent.

"Part of the urgency got sidetracked into the nuclear power debate," Kovarik says. "People say it's about capitalism but it's about a failure of leadership. . . . When young people look up and say, 'Where the hell did everybody go and what's going on?' I don't blame them."

Gen Z. Where does that leave them?

I'll let this chilling paragraph surmise. It's from a white paper about Gen Z's flare for ironic humor, a deep dive into things like young women mocking feminism with knee socks and anti-immigration T-shirts on social media: "Underlying the emergence of youth reactionary culture is the glaring failure of neoliberal capitalism to deliver on its promises. Young Americans will face the worst economic odds of any post-war generation. Life expectancy continues to drop (mostly in Red states). Teenagers cannot identify the causes of this general decline but they have already internalized its downward trajectory. Their grandparents' single-earner income provided for a four-person household. Their parents' same family unit now struggles to reproduce its standing with two earners in the workforce. Just a few years ahead of them, Millennials with multiple roommates all tweet about the perils of student debt and freelance precarity. These material considerations, combined with the real existential threat of climate change, make Gen Z acutely aware that they are living at the end of an era; that they were born on the other side of the so-called 'end of history.'"

It's all just the system, right? It's always been the system.

"Gotta have faith in the system!" says Andrea Rodgers. She was among the next attorneys to join the *Juliana* legal team, and this comment, this is sarcasm. She has the guilt too.

Rodgers has long had faith in the law, though. Legal aptitude runs in her family. Her father, also a lawyer, filed the first petition to ban DDT, sued the chief of police of Seattle, and spent a career fighting the corporate takeover of the American government. He also represented the Puyallup Tribe in the Fish Wars, among the first cases to put Northwest courts at the forefront of exactly the kind of judicial remedies the Juliana plaintiffs are seeking: one in which a judge would put executive agencies under supervision, in that case for failing to enforce treaty rights to fish. So Rodgers grew up understanding that fighting for what's right is just what you do. And that the environment is almost always in peril, no matter how much you do it. She was taught that courts are for justice. And that when there's an injustice, the courts are there to fix it.

Still, after the trial is canceled, she worries about how slowly the courts have responded to the youth. "It's been . . . years since we filed our complaint and they still haven't gotten to have their case heard at trial, and I'm asking them to believe and have faith in that system. They've been plaintiffs, some of them, for a third of their lives," she says. "I think, for them, that's hard to do because they have been in the system and have seen what it's like."

Olson wants her clients to have the chance to talk about how this weighs, how bruising it all is. Somewhere public. Somewhere that matters. It's the kind of experience she wanted as a victim of gun violence but never had. "I had a lot of help healing physically. But emotionally, I was not supported. Friends didn't understand, family, teachers," she says.

The afternoon light is sliding through the window and she says there was a neighbor—a quiet man, a Vietnam veteran with older children—who seemed the only one to see her then. He had a Purple Heart from the war, and he came to her home and sat with her one afternoon. Gave her the Purple Heart. A powerful moment at a time when it was a struggle to connect.

Now there are support services for such things. More mass shootings, more people who live in communities that deal daily with gun violence, more public awareness, more places for victims to go, for youth victims to stand up, to advocate for safety, to say, *This is how this feels.*

"God, if my plaintiffs can ever sit on a witness stand and one day do that, that will be huge. Giving that voice to youth and that ability to really participate in these existential issues is so important."

Olson makes this association between the trauma of gun violence and the trauma of environmental collapse from the unique vantage of having seen both. And because she understands that trauma is just trauma, whether it hits you in the form of a bullet or in some other way.

CHAPTER NINE

November 2018: Doubling Down on Backward

November 2018 begins a period Nick will later refer to as "not the happy times, to put it lightly."

Trial had always seemed precarious. But when the Supreme Court told the lower court to go ahead in July, it seemed like the last hurdle had been cleared. A month ago, before the trial, Nick was looking forward to the days when scientists would finally testify, finally tell the courts about all the harm climate change had wrought—would still bring. It was what he had wanted all along.

Now, in retrospect it seemed like the government was working overtime to keep the trial in a theoretical phase, the one in which the case was just a judicial overreach by a bunch of well-meaning kids. That way, Nick says, the government could pretend the merits of the case were good, the spirit valid. It was the legal nuts and bolts that were objected to. This sought-after remediation plan? Simply not the best way to get it done.

"That kind of denies the real thing of, 'This is going to drastically screw over future generations,'" Nick says on the phone, back in Colorado, in cooler climes, finally, but with trial canceled, life rearranged.

The government's position strikes him as ironic. "This case is the only way to get it done because every other effort has failed, essentially."

Still, the trial could resume anytime. So Nick begins the waiting, along with the rest of the plaintiffs.

The world turns through this waiting, but strangely. To the south, a caravan of refugees marches toward America's southern border, its members propelled at times by crop shortages and slumping economies after a series of weather-related calamities and drought. In response, Trump calls on troops to defend the perimeter and issues campaign ads so full of hate and rhetoric that even Fox News takes a stand against airing them. On the campaign trail, Trump makes comments about terrorists and extremists lurking in the masses. He isn't running for re-election, just control of the House and Senate by his party. He is running on the platform of hate and fear that seems to galvanize his base, though it sets another kind of American reeling, the kind that is turning out en masse to vote away the anxiety in the midterms.

They are Americans who want "the world's greatest democracy" to still be that and who trust that by voting, they will set right what seems to have gone off the rails. But pining for an American ideal runs very far behind the despondency that is setting in for some of the young. So caught up in the daily polarization of it—the self-righteous defending of the guns and the identity politics, the mad rants on social media and the friending and the unfriending—many of the adults are missing the facts that Earth itself is destabilizing and that the youth are getting wide-eyed.

A few freshman congressional contenders see this and start saying the words "climate change" with something other than a tone of taking attendance, a ticking off of things in a rote roll call. The ears of the youth, they pique, turn. Something begins to shift.

Still, while people with jobs are sleeping in cars and medical bills are towering, swallowing whole families, at least one Canadian news outlet is forecasting an American civil war. It is hard for some to see that Earth is breaking down when our nation is breaking down so grandly, so radically, day by day in newscasts. As the hate intensifies, there is an attack on black shoppers at a Kentucky Kroger by a man who'd hoped to stage an assault on a black church but failed. Days earlier, another man had been arrested for sending pipe bombs to some of Trump's favorite targets—people who'd criticized the president, including media—in south Florida, where the man had been living in a van plastered with Trumpobilia and hate speech.

It is all very distracting. And the headlines are fantastically good cover for the broad deregulation of environmental protections that is swiftly under way. When he is not lambasting his detractors on Twitter or the campaign trail, casually threatening wars, firing people, or lauding his own achievements as the leader of the free world, Trump is steeped in an effort to undo the rules that stand in the way of corporations. Despite the risk to youth, what it spells for an America the president will never see, his administration doubles down on moving the nation backward on ecological gains. Before November 2018, Trump had already rolled back protections for endangered species, signaled plans to pull out of the Paris Agreement, dropped climate change as a national security threat, cut the NASA Carbon Monitoring System, and cut climate and clean energy programs in a proposed budget. In the days after the almost-trial, his regime is rewinding fuel efficiency standards and promising to strip California of the ability to impose its own. It doesn't matter that even the auto industry is against this, having been thrown into disorder by the sheer impulsiveness of it.

Ironically, the Environmental Protection Agency is leading the Trump administration's assault on the environment. It's been a year since the agency removed the informational page on climate change from its website, presumably for an update that never updated, prompting one former official to call it "the ghost page." In between, the EPA has repealed Obama-era rules curbing methane and emissions from coal plants, and loosened regulations on air pollution from toxics—including benzene and asbestos. It has also disbanded its advisory panel on air pollution, approved the spraying of a pesticide (chlorpyrifos) scientists say causes brain damage in children, and tried to delay a public health report on deadly industrial chemicals (polyfluoroalkyl substances, or PFAS) as they seeped into the water.

As November begins, the EPA's press shop is being led by a Republican political consultant, and its staff is getting bare-knuckled with reporters, calling out alleged errors in press releases and blocking some members of the media from attending events. The president has already made his "enemy of the people" comments about the press and stripped CNN reporter Jim Acosta of press credentials for being bombastic. The EPA appears to be following suit, not only blocking reporters from attending things but also publicly calling one "a piece of trash" and another "an anti-Trump reporter" and "dishonest." Suddenly even access to information about the environment demands loyalty. Hard to have when the EPA's Region 4 director—former coal lobbyist Trey Glenn—is out of jail on bond for helping his former employer dodge a Superfund cleanup in Birmingham. In short order, the scandal-plagued Scott Pruitt, whose spendthrift ways are so lavish he's managed to fritter away $43,000 on a phone booth, will be succeeded as EPA administrator by Andrew Wheeler, a coal and energy lobbyist and a notorious zealot for deregulation. As an interim director, Wheeler has already proposed rules to loosen carbon limits on power plants and change the way the EPA is calculating the benefits of air pollution standards.

This wholesale reversal of environmental ethos spreads beyond the EPA. Ryan Zinke, secretary of the interior, is among its proponents. In the early days of November, the Department of Justice is investigating whether Zinke enticed oil megacorp Halliburton to build him a brewery while the company lay in wait for Montana public lands to open for drilling. Also pending are more than a dozen other inquiries into Zinke's exploits, including attempts to exempt Florida from offshore drilling rules, censoring a climate report, and killing a study on the health impacts of blowing the tops off mountains in Appalachia to mine coal. The Department of the Interior has already rolled back protections for water, wildlife, and air to make way for energy development and is offering more than 12 million acres of new federal land for lease. Soon, Marcy Rockman, who is charged with studying the effects of climate change on the nation's parks, buildings, and archeological sites, will post her resignation from the National Park Service on Twitter, essentially telling the world that she is being starved of the resources to do her work. Somewhere in between, Steven Chancellor, the Indiana coal mogul who'd been tapped to advise Zinke on hunting, secures the permits to import lion heads into the States from Africa.

Elsewhere in the executive branch, the Forest Service is distracted from the business of managing the forests, battling instead "fresh revelations of rampant discrimination, bullying, retaliation and sexual misconduct," according to the Associated Press, even as scientists discover that the last of the world's wildernesses are disappearing and that America is one of five nations stewarding the bulk of them. FEMA's maps, it is revealed, are years out of date and failing to account for sea level rise and extreme weather. The Departments of Energy and Agriculture are promoting the burning of trees and other biomass for energy, even as scientists warn the practice could be more damaging to the atmosphere than burning coal. And an analysis shows the Pipeline and Hazardous Materials Safety Administration is failing to fine pipeline companies more than 90 percent of the time when pipelines explode or catch fire. Meanwhile, the Coast Guard is making changes in its still-failing efforts to contain a fourteen-year-old leak in the Taylor Energy oil well off the coast of Louisiana that dumps roughly 700 barrels of oil into the ocean every day.

Back in Louisiana, Jayden calls fellow plaintiff Miko on the phone to rant. She's supposed to be at trial, supposed to be with the other plaintiffs in Oregon, but within a week she has been boomeranged back to the place she'd been battling in all along. Lately her mind is drifting to some not-so-good thoughts. Asked to describe them, the first word she utters is just "Oof." She says she has a feeling of momentary hopelessness.

"I love the case, I love being part of the case, I love all the plaintiffs, and I love all the staff, but, like, *come on*," she says. These fits and starts, she feels toyed with. "Having to have all this excitement just to go home empty-handed is something that would get everybody angry," she says. "When I think about it, I'm just like, man, *what* is going on?"

It doesn't make her lose hope. Not exactly. "But it kind of did put me a step down in a way," she says.

It feels like the government is purposely stalling, purposely making the plaintiffs wait. As if "we need to go back to our place," Jayden says. "It's kind of showing that they don't really care, you know? So yeah, I do kind of take that personal. Because you can't even face twenty-one kids in court? How does that make you sound any type of good right now?"

She says she isn't mad at everyone. Just rich people, mostly. The ones who put their money into destroying the things so many others need. The ones that invest in fracking. In drilling the Gulf. In making the extraction market so, so big that the damage belongs to everyone now.

As November inches on, politicians do as politicians do and make loud plans to fix it all in the last days before the general election. But even as oil companies dump millions into state races to fight carbon taxes and oil-and-gas zoning, there are signs that the environment is winning—with the people at least—as Americans step to the ballot box. Republicans in close races are talking climate change, for example in Florida, where congressional incumbent Carlos Curbelo floats a carbon tax proposal and state Republicans are torpedoing his Democratic challenger for taking "dirty coal money." Governors' races in some states also show a likely push toward local regulation of the very things the federal government is presently deregulating—like air quality. Some candidates are talking renewable energy mandates. And on the campaign trail, the language of the Green New Deal begins to foment, to take shape as more than a pipe dream while a contingent of Democratic up-and-comers urging radical spending on climate change look to be front-runners.

They win on November 6, and Democrats take control of the House. But to think that this changes much would be neoliberal, absurd. Party control by Democrats can slow but not halt the assault on Earth. It is the young—soon to stage a protest in House Speaker Nancy Pelosi's office—who can most clearly see what many older Americans cannot: that the problem of the collapsing planet needs more than this biennial commotion of voting for the people who might care. Or voting people out of office who don't. Whether saving Earth is becoming synonymous with dragging politics to the left doesn't matter. The work of doing so needs to stop falling second to simply winning, as if all that needs doing is cheering in some kind of biennial horse race. This national delusion that blue victories are enough is a diversion. The problem demands a focus on things our nation is not used to even seeing.

As if to make this point, the *New York Times* reports that week that Democrats have flipped twenty-nine seats in Congress, with average districts shifting 10 percentage points to the left—"the crest," it posits, of a blue wave. In later print, though, the *Times* notes that the wave is only half the size of the red wave of 2010, during which districts shifted more than 19 percentage points to the right. In other words, the nation's electorate is wresting control of itself from itself. The news that it has shuffled back across the ledger is really only that.

In the case, meanwhile, there is a brief flicker of hope. On November 2, both the Ninth Circuit and the Supreme Court finally deny the government's applications to stay the trial, in effect paving the way for it to move forward once again. But on November 5, the day before the election, the Department of Justice files two more briefs in the *Juliana* case—one with the Ninth Circuit Court of Appeals and the other with the US District Court of Oregon. The first—another emergency petition—is aimed at killing the case yet again, the second at stalling it while this new request for dismissal is considered.

The legal team for the *Juliana* plaintiffs is exasperated. "It's unheard of," attorney Phil Gregory says in a statement. "I've asked colleagues, former Supreme Court law clerks, we've asked legal scholars, reporters who cover the courts, and no one has seen this before. It's the clearest evidence of how fearful this administration is of standing trial in this case." He says it's the fourth time the government has made these arguments to the Ninth Circuit Court alone, and each time the plaintiffs have been made to wait. "It's outrageous and shows such disrespect for the process."

No bother. On November 8, the Ninth Circuit once again stays the trial. It seems the government's attorneys can do this forever, as long as the courts will let them: ask the same questions over and over, and use the upper courts to pressure the lower ones, meanwhile avoiding a trial. But the courts are the courts. Procedural. Careful. Any claim of misstep, of error, is carefully assessed instead of treated as a symptom of what it is becoming: obstruction. I send the news to my Reuters editor, who cheekily dubs it another episode in the legal soap opera that has replaced our courtroom thriller. "As the World Burns," he calls it. In the sick way that gallows humor staves off frustration, we laugh. What else is there?

Later that day, though, the world does burn, and the joke is no longer funny. Not one but two major wildfires erupt in California, killing scores, decimating a small town and celebrity mansions alike, burning buildings to ash and driving hundreds of thousands of people from their homes. Drought conditions had been an issue, the unceasing heat an invitation to tinder, but despite such signs that wildfires are worsening, no word of a trial comes.

While observers of the *Juliana* case wait for things to go back to normal or settle into a new one, the City Club of Eugene hosts a panel discussion about what to expect next. There is, of course, nothing to expect. But conjecture is still healthy for supporters, who by then have such a case of whiplash over the trial starts and stops that about 150 of them turn up looking for clarity or just a few more kernels of news to egg on their fanship.

One woman shows up in a baseball shirt emblazoned with "Trial of the Century." A man notes on his name tag that he has "arrived by bicycle." People do what they can.

Alex, like Avery, saves his anger for himself. Back in New York, he is disillusioned, annoyed at having gotten his hopes up for trial.

Kiran is back at college, too, and trying not to worry about what happens next. There is still one semester left. But this is the last semester that really matters, Kiran having arranged the spring around playful electives to ease the transition to slackerdom. Worries are, for now, dancing on the other side of an anxiety wall in the push toward winter finals, a wall that will fall in another month.

Meanwhile Aji Piper, also in Seattle, is wrestling with the pretrial image of himself. Scholastically out of the box for roughly half a year, Aji ended his high school career in the spring of 2018, not having received a diploma. Having endured the pretrial coverage typecast as "a high school dropout," however, he is disgusted with this image of himself and resolves to reenroll for the one last hideous semester.

Levi, ever the optimist, concentrates on keeping his room clean. It is not usually this way—clean—but he was tasked with scrubbing it before he left for trial, and after returning to Florida finds that he likes it.

"It was actually kind of nice to . . . be able to keep it that way for a while," he says. And he enjoys focusing on this tidiness, on exerting control over his own dominion.

It was upsetting to have the trial canceled, he says. And to have to do things like turn up at the pet sitter's to pick up his hermit crab early and explain to the people from church—the ones who had only just thrown him a going-away party—why he was home again, especially to those who were not so acquainted with the Internet. The time he had with his co-plaintiffs, without the courtroom, without the drama—he remembers those parts well. The mushroom festival and the hangouts. And he is glad for all the people who showed up to rally who had never been plaintiffs but were as upset as plaintiffs anyway.

His feelings toward his government when the trial got canceled, though, were not so positive.

"It just felt like they didn't care about us enough to let us have a trial."

As the month wears on and the waiting continues, bad news about the state of the natural world rains down in a cascade. The World Wildlife Fund (WWF) delivers the news that people have wiped out 60 percent of the animals on the planet since the 1970s. And as if the coral reefs aren't having a rough enough go of it, scientists learn they are starving for nitrogen off islands where rats have infiltrated, the rats having eaten the birds and island bird poop being, it turns out, a mainstay of nitrogen supply for coral reefs.

Elsewhere the rats, like the raccoons, are proliferating in the heat, and scientists at Cornell University warn of a ratpocalypse. Rats can breed within a month of being born and have a gestational period of fourteen days, meaning a rat can be a grandparent by the time it is three months old. The higher temperatures are spelling fertile breeding, and scientists predict cities could be overrun by rats as the world gets warmer. The implied caution is that not all of the rats will be Disney rats, famous chefs like Remy in *Ratatouille* or good with wardrobe like the rats in *Cinderella*. Instead, rats can be prolific carriers of disease. In a roundup of rat news, the *Boston Globe* further reports they are playing in puddles by the dozen in Boston, and tenants in a public housing project in New York say the rats are getting bigger and have lost their fear of humans.

In this way, it seems the animals are either winning or losing in the climate fight, mostly losing.

In the losers column, Atlantic waters are getting so warm that shellfish are on the decline, prompting experts at the National Oceanic and Atmospheric Administration to predict that the futures of scallops, quahogs, softshell clams, and eastern oysters lie in farming. That outlook seems optimistic, though, after farmed oysters in Florida, already pummeled by Hurricane Michael, are then made central to a water war. Also, heat waves are causing infertility in male insects. The butterflies are having a difficult run, too, with warmer temperatures upsetting migration and plants, while habitat for more than a hundred species at the National Butterfly Center lies in the unfortunate path of Trump's sought-after border wall. The possible fate of butterflies? Cast to farther-flung lands or, worse, immortalized in glass frames and hipster tattoos. The Bureau of Land Management is selling off the habitat of threatened sage grouse. And an insidious wasting disease is infecting deer and elk in Wisconsin.

It is hard to define the situation with the shorebirds. Plovers and sandpipers and the like, about half of which are already in decline, are fending off a threefold increase in nest robbing by Arctic foxes and weasels and such, likely because global warming is making eggs a best-bet for food and thinning grasses make it easier to find them. Many of the affected birds are cute and fuzzy, with offspring that make for greeting cards with grammatically mangled salutations like "Seas the Day" and quotes from Proust: "The real voyage of discovery consists not in seeking new landscapes but in having new eyes."

Reading it all, my eyes are killing me.

On November 21, three weeks after trial was canceled, Judge Aiken certifies *Juliana v. United States* for interlocutory appeal, inviting the Ninth Circuit Court to inspect the case, halt it if it wants to. It's a dare, it seems: You want the case? Take it. As she does this, Aiken writes that she thinks *Juliana* is better off in a trial court. Her trial court. But she asks the Ninth Circuit to have a look at the case if it wants to. Observers will later note that Aiken seems to have been pressured to do this. That in its first denial of the government's petition in July, the Supreme Court justices had hinted in unsubtle ways that the case ought not to go to trial, given that "the breadth" of the plaintiffs' claims was striking and presented "substantial grounds for difference of opinion." They had urged Aiken's court to take the government's concerns about "the burdens of discovery and trial" into account.

Aiken had let the case go forward despite this language. But subsequent rulings from the Supreme Court and also the Ninth Circuit Court seemed to urge her to let the Ninth Circuit judges intervene.

Stephen Vladeck, a professor at the University of Texas School of Law, will eventually dissect all of this in an essay for the *Harvard Law Review*. He will put it this way: "In a superficial sense, the government 'lost' its requests for emergency and extraordinary relief in *Juliana*.... But ... still nudged the lower courts to provide much of the relief the government had sought."

The day of Aiken's ruling, however, following what is now a typical Trumpian rant about the Ninth Circuit and "Clinton judges," Chief Justice Roberts takes the rare step of firing off some choice words at a speech. He says there are no Clinton judges, no Bush judges, just judges. But despite this assertion of judicial independence, observers like Vladeck will conclude that the court has been bending to the administration's tactics while it caters to all these "emergencies." The plaintiffs' attorneys are still watching several other cases, watching the Department of Justice run roughshod over the courts with the same unprecedented calls for relief: still targeting the ban on military service for transgendered people, the rights of young Dreamers to remain in America, and a citizenship question on the US Census.

That night, I watch a three-part dystopian Western called *Young Ones*, in which Michael Shannon stars as a besieged family man who defends a muddy well, a job, and his children in times of epic drought. The world is a dust bowl and there are Spaghetti-style close-ups against barren, useless landscapes. Between scenes, the camera pans quickly across sand. Against this desolate backdrop, people are somehow still driving. They drink water from gasoline pumps, and the lure of the land is a man who hawks machines at auction, drawing its sparse inhabitants, zombielike, by calling for bidders over the airwaves.

Around this time, my husband and I pass a night with two friends who have returned to America from India not long before. They are next headed to Canada, unwilling to wait for whatever comes next in our nation's continuing dissolution. Our friend has just written a book about Silicon Valley, one that uncovered surprising connections between the valley's elite and the rebirth of white supremacy in America. A few days before in Oregon, the state's leading newspaper had run an editorial sympathizing with a Patriot Prayer patriarch who'd been using our community as a performance space for a nativist agenda. My friend was the only journalist in the state to speak out against this pandering. He is worried about what that means.

The coffee table is lined with prescription bottles. He says the landlord doesn't allow pets, but he and his wife have a letter from their therapist endorsing their kitten, who they pass the hours teasing with a string on a stick. In a few days, they will load the cat into a U-Haul and be off to Edmonton, where they fear the cold and the tar-sand politics but feel safe, at least, from what they increasingly view as the impending civil war the Canadian press has predicted. People sleep in the streets. The richest among us control a government that is gutting public services, calling them entitlements. And so many guns, many in the hands of people who seem increasingly willing to use them against those who do not agree with their views or look like them. Now this: the courts are breaking.

We talk long and mostly drink while I shop from their cast-off clothes. The later it gets, the more paranoid we all sound. Talk to any investigative journalist, my friend says, and you'll start to wonder if they're off their meds. Another colleague is writing a book about plastics, and all she can talk about is how we are all slowly becoming plastic. How scientists have found microplastics in people's bodies, in their excrement.

It is a race to what will kill us first, it seems: the violence we are careening toward, our garbage, or our dying planet. Days later, when scientists posit that an elongated object found in space might be an alien spacecraft, my friend shares the news on Facebook, delighted. "Help is on the way!" he writes.

I soon take a walk in the woods with Isaac and Miko Vergun, siblings and *Juliana* plaintiffs, the only sibling duo in the case. Isaac is sixteen and, on most days, a study in fashion. The clean lines of his suits, sharp collars, and the glint of his glasses bespeak a precision of dress and a certain whimsy too. Bright colors. No fear of patterns. Today he turns up in a beige pullover over a safari-style button-down, the front decorated with embroidered critters. There's a bird with a watering can with a heart on it, watering a daisy in a plaid pot; a chicken with a blank expression; a spade with garden gloves; and a garden stake with radishes on it. He is wearing checkered shoes.

Isaac knows there are solutions to climate breakdown. And he's made a certain peace with humanity's slow course in finding them. While we talk, he tells me he expects to live in an apocalyptic world someday. "Probably when we're like thirty or so, it's going to get kind of bad," he says. "But like at that point, hopefully, people can put aside their money interests and then realize that we're actually in a bad situation and then all work together. I'm obviously trying to make that happen sooner."

Isaac says this with no anxiety, no bitterness. Just the calm, matter-of-fact way of a person saying what's so. He thinks there will be more smog, more fire, more hurricanes until people are motivated to turn back.

CHAPTER TEN

Crashing the Official Narrative

We are still expecting a trial. And the press, like the plaintiffs, is still thinking it might be soon. The planning is delayed but not over. Thus the usual jockeying for position in the courtroom carries on.

Members of the press had already complained about the lack of priority seating at the would-be trial in October, worried about arriving with the public at 6:30 a.m., standing in line for seats, and never being able to leave without yielding valuable courtroom real estate. So early on, Judge Aiken made a decision: since this would not be a jury trial, she would seat the press in the jury box.

Fitting, in ways, for the media to sit broadside attorneys for a president who had called them "the enemy of the American people," cried "fake news" at unflattering facts, made network affiliation a popularity contest, and under whose leadership both physical attacks against journalists and attacks against press freedom had ballooned.

The relationship between the American press and the White House had rarely been worse. But then, the relationship between the American press and the public had also rarely been worse, barely rebounding since public trust in media hit the all-time low of 32 percent in 2016. Entertainment and news divisions at television networks had been battling each other for decades, and by the first decade of the twenty-first century, entertainment had clearly won. News now spilled from a television universe in which journalists were flanked by pundits, and predictable audiences gravitated toward the loudest of them. Corporate hawks secured a stranglehold not on talent but on numbers: predictable audiences meant predictable ad rates, which secured predictable cash flows. As seasoned, qualified news anchors became surrounded by personalities, the nation counted on these people for fun as well as facts. The networks, meanwhile, blurred the lines.

Now, ask a climate denier why they don't believe in climate change, and what you'll get is talking points from *Fox & Friends*. That's our fault, in the fourth estate. We can blame Trump for being Trump, past presidents for their inaction, and politicians and pundits for politicizing climate change. But as members of the press, we also have to blame ourselves for fueling debates about climate science rather than conveying the facts. The seed of denial was planted by propagandists. But it was the media that legitimized it and overlooked the outcomes inaction would spell for youth.

Xiuhtezcatl has been uniquely entwined with this problem. He grew up in the middle of it, most of his years spent with a camera pointed at him. Before he was sporting the fancy Adidas, had his own billboards and a couple of albums, too, he was just a small person making heartfelt speeches, sometimes to large crowds, an anomaly as activists go.

He would say in his book, *We Rise*, that he heard this call to protect the planet early on in life. That as an Indigenous person, he had always felt a connection to this duty through the ceremonies and dances of his Mexica people, and through the stories his father sang to him when he was a child. He says the tale of his ancestors, the story of the Aztec Nation that came before him, is the story that carries him forward. He sees his life as part of a longer chain and sees himself as one person in a long line of people who fight for the earth. He wrote that his long hair, a life lived in ceremony and danced through deserts, frequently remind him of this role. So too his name, chosen by his grandfather and other elders based on a study of the Mexica calendar. It means, in rough translation, "blue turquoise mirror," a reflection of the sky and stars. His middle name, Tonatiuh, references the sun, his shine.

Xiuhtezcatl would grow up to be a media savant. After roughly a decade and a half on the other side of the interview, he is as qualified as anyone to talk about the stories the media missed while it was ceding airtime to pundits, somewhat deaf to the rising concerns of youth.

I am eager to ask him what this was like, but it takes me about a year to get an interview with him, to persist to the front of the line. By the time I get him on the phone, we are in a post-Greta world and Xiuhtezcatl is commanding a hefty speaking fee for crowds, his activist career having segued into hip-hop notoriety and fashion modeling. He's working on a third album, has just collaborated on a single with Jaden Smith (actor Will Smith's son), and recently wrapped up a grueling national tour. He's nineteen when we talk, and this road for him has been long.

A year after *An Inconvenient Truth* mainstreamed the climate conversation, when George W. Bush was still president, Xiuhtezcatl was a seven-year-old who literally ran fire all the way from the Mexican border to Hopiland in Arizona in a prayer ceremony aimed at "strengthening our commitment to protecting the sacred elements that give us life." He described in his book how he saw Leonardo DiCaprio's documentary *The 11th Hour* then, and how it affected him horribly. The love of nature he'd inherited from his father's people began to mix with his mother's activism. Tamara Roske was the co-founder and director of Earth Guardians then, which began as an alternative high school in Maui and is today an international conservation organization focused on youth leadership. Xiuhtezcatl became the organization's youth director, led direct action campaigns to challenge pesticide use and the spreading of coal ash in his hometown of Boulder, Colorado, and helped to build an international network of youth activists in the fight against climate change. When he was fifteen, he became only the second person not affiliated with government to ever address the General Assembly of the United Nations, something he did in three languages—English, his native Spanish, and his ancestors' Nahuatl.

As the climate coverage swirled around him and his own advocacy took flight, Xiuhtezcatl says he was tokenized but not seen.

"When I was younger, I think media outlets were more like, 'Well, this is confusing, why is this little kid out here doing this work?'" He got the sense that journalists didn't know what to do with him but chose to report on him anyway because he was something out of the ordinary, a novelty. They quoted his words, his urgency, his résumé, but missed the story of what his presence in the climate fight meant, how many young people shared his concerns. So while the media paid attention to him, Xiuhtezcatl remained someone whose story was not always clear, even while his ubiquity slowly became a fact of the environmental times. The larger story of youth activism was also similarly overlooked, left behind the curtain of official things that journalists paid attention to. This massive generation that would stand up with Xiuhtezcatl by the spring of 2019, they didn't wake up to this cause overnight. Their ascendance was happening all along, a thing unseen.

To understand this media blind spot, consider that to be a journalist is to be a slave to your editor and the publication she rides in on. And it's important to know that most editors do not get that way by spending years on the science beat. Writer Ross Gelbspan described it this way, in his book *Boiling Point*: "The career path to the top at news outlets normally lies in following the track of political reporting. Top editors tend to see all issues through a political lens." They work their way up as political correspondents, most starting out covering local or state policy and climbing to higher levels of government with increasingly complex problems to solve. Their bread and butter are issues that are subject to debate. They speak a language of points and counterpoints. About the right way to tax a thing. About whether one type of road striping works better than another. All of it is a matter of at least some subjectivity, of opinion, and to become too confident in a particular side is to lose one's objectivity, to be mentally flaccid, to have lost it.

These were easy claims, then, to make against the science journalists for whom facts about climate change were not in dispute. Covering global warming was a lot like covering a fire or a car crash—there was little arguing about events that had taken place, and unless there were mysteries, of timing, of inches, of lives still in the balance, there was little in dispute. But for as long as there were people insisting on another set of facts, on debate and opposing views, the instinct to balance those viewpoints would be exploited full tilt.

And there were, indeed, people insisting on another set of facts. Since the first Bush presidency, with its deep ties to the oil industry, fossil fuel companies had been setting a grassroots backfire despite mounting evidence that burning fossil fuels would lead to a planetary crisis, as intellectual and environmental observers continued to sound the alarm. Rather than adapt, the industry began seeding dozens of little think tanks. The Heartland Institute. The Cornwall Alliance.

Where formerly there was a basic set of agreed-upon facts, if disagreement about solutions, suddenly "there's a crazy quilt of right-wing mouthpieces that all come really out of the oil industry. And they're saying crazy things like 'the moral case for fossil fuels . . . ,'" says historian Bill Kovarik. "In the old days, these things were hashed out at the treetops level. . . . You didn't have people who didn't know the first thing about science trying to argue that the radiant force of equations was wrong or that Michael Mann's hockey stick was flawed in ways that they couldn't even begin to describe but they just heard that somebody else said it was flawed."

By 2000, when the second Bush presidency was under way, the United States had begun to break with international allies by turning away from the climate reality, growing publicly skeptical about evidence of warming. George W. Bush appointed a cabinet with close ties to the fossil fuel industry and adopted disinformation as an official strategy. Information about climate change was also edited out of an EPA report. And a campaign promise to cap coal emissions was reversed. Instead, the next energy plan called for more oil exploration and thousands of new energy plants, many of them coal. And then the ouster of the IPCC chair happened at the request of ExxonMobil in a bid to replace him with a climate skeptic.

This was the year Xiuhtezcatl was born—at the dawn of the age of official denial. And the emerging polarization that became a backdrop to his personal story—a presidential administration aligned with corporate astroturfing, banishing science—was a tightrope walk for the press too.

"The nature of journalism is that we were all trained to get multiple sides of the story and so we did our best to be balanced, and that was kind of the standard at the time," says Jim Detjen, the former director of the Knight Center for Environmental Journalism at Michigan State University. He used to be a reporter and an editor. "If you were working at a mainstream paper there was, I think, a lot of attention to the balance because you were criticized by people from the oil industry or fossil fuel industry for your coverage."

Detjen knew how intense this pressure could get. He'd reported on Three Mile Island for the *Philadelphia Inquirer* after its partial reactor meltdown, the one that caused a radiation leak near Harrisburg, Pennsylvania, in 1979 and became the worst nuclear disaster in US history. Detjen and another reporter wrote a series in the early eighties critiquing the fourteen-year cleanup. At the time, the utility in charge of this mess had a sophisticated public relations office tasked with making it smell like roses. They took out two-page ads in the Sunday paper roughly calling Detjen's reporting a hatchet job, then held press conferences denouncing him and threatened lawsuits. Climate change coverage was attended by these same accoutrements, but everywhere and daily. There seemed little limit to the resources the fossil fuel industry could apply to leaning on newspapers as newspapers took the lead in covering the environment. Phone calls to top editors. Meetings some reporters suspected they were not privy to. These things were the stick.

The carrot looked more like this: "I remember getting a videotape in the mail . . . that was all about climate change, and then the focus was: climate change is going to be wonderful!" Detjen says. It claimed climate change would cause plants to grow faster. And that more carbon dioxide was a good thing, practically converting the planet into a utopian biodome. When he began teaching at Michigan State, Detjen used this video to teach student journalists what kind of propaganda they were up against.

To ask reporters from the era, many editors did well to swat away these carrots, but they had a harder time with the sticks. With the phone calls and the meetings. The industry reps crying foul and charging reporters with the cardinal sin of having lost objectivity. Reporters were always, in the end, just employees of the institutions they worked for. They did their best to play by the rules. To be cautious about what things were fact, what was still in debate. Scientists did not help much. They were busy being scientists instead of standing up to correct things. And the meteorologists the public relied on to talk about the weather, many of them were among the skeptics.

This instinct to be fair and balanced, it's not a bad thing. "Skepticism is one of the greatest virtues in a reporter. The most important thing a reporter can ask is 'How do you know that?'" says Joe Davis, who covered environment and energy for *Congressional Quarterly* through the early days of climate denial. "That's all well and good, but the purveyors of disinformation know that that is part of a journalist's ethic and they use it and manipulate it."

Not a few of these journalists talked among themselves then about whether this was the right way to cover climate change. Several had founded the Society of Environmental Journalists so reporters tasked with covering global warming and other environmental matters could help each other do the job. They talked, not infrequently, about whether balance was appropriate when it came to scientific fact. About whether they should be more strident in crushing this idea of debate. But they also wanted to keep their jobs. Climate change was still, as Gelbspan would describe it, "ghettoized as a sub-beat of environmental reporters" rather than integrated into coverage of policy, politics, business, and the other issues that seemed to matter more. No one wanted a personal story like that of Phil Shabecoff, the former reporter for the *New York Times.*

"I have no direct evidence of why I was taken off the environmental beat. I was told my coverage was too alarmist," Shabecoff says of his 1999 departure from the paper after twelve years covering the environment.

In his tenure at the paper, Shabecoff had moved the *Times*'s coverage of global warming from the back of the paper to the front. He'd heard complaints from industry sources along the way. And he would play the balance game, he would report both sides. But every time the other side was in the employ of the fossil fuel industry, he reported that too; noted where they drew their paychecks.

"I had been supported by the top editors of the paper . . . but there were new editors at the time. And the national news editor had spent most of her career in business news." He surmises industry reps had reached top editors with their complaints, but he could never be sure. "I was taken off the beat because my stories about climate change, loss of species, and toxification of the environment were considered too alarmist. In retrospect, I wish I had been more alarmist."

He left the paper and founded Greenwire, making him one of the first journalists to build new media—a new water cooler—for an issue that was being pushed, aggressively and at great ethical expense, to the fringes of the national conversation.

I'll let Xiuhtezcatl describe how all this looks to someone born in the middle of it, and how it appeared to him throughout his activist life: "It didn't feel like we were taken super seriously. There was excitement around young people being involved, but again it was much less than there is now," he says. Journalists didn't dig deep on the urgency of climate action for youth, didn't ask why the youth voice was rising, didn't report on what those rising voices were saying about intergenerational equity in a world on a relentless consumptive bent, pushing ecology to an extreme.

Since then, "there's been a massive shift." Social media raised the voice of the affected generation finally, Xiuhtezcatl says. And it became the chief organizing tool of young leaders across a spectrum of social issues: climate justice, gun violence, Black Lives Matter. All used social media to transform the way their stories were being told. This changed the way people perceive youth leadership in America. In the world. Suddenly reporters tuned in. And media organizations began taking note as youth calls to address the climate crisis rang out with increasing urgency.

"Because of social, young people have been able to crash the narrative more on our own without depending on mainstream media," Xiuhtezcatl says. Editors would agree, the *Columbia Journalism Review* (CJR) writing that social media allowed youth activists to begin driving the news, initiating actions and protests that called media to the cause. They wrote op-eds, produced videos for YouTube, Vice, and Vox, and posted and responded to climate news with enough fervor to keep the conversation spinning around the clock. *Teen Vogue* editor Lucy Diavolo would summarize, telling CJR, "Social media has altered the landscape of authority." Even still, press coverage remains fuzzy on why youth can claim this authority—if not disbelieving, then at times failing to acknowledge the degree to which the next generation's inheritance is spoiled.

To speak these truths, any way he can, is more in line with the way Xiuhtezcatl grew up. With empowered siblings and fellow activists who also fought for the earth, other children who ran fire across desert, who climbed to mesas. Kids who danced prayers until they could not. For them, their identities were invisible in all kinds of ways. And their media identity remains somewhat indistinct. Google Xiuhtezcatl and you'll find featured snippets of his Wikipedia entry; channel recommendations for YouTube, Spotify, Pandora; links to his Twitter, Instagram, his Facebook. It's the digital life of a person who lives loudly online but whose presence the mainstream press still struggles to understand.

"They just have failed so much in the last ten years at really capturing the energy and the essence of what stories need to be shared" about the youth movement, Xiuhtezcatl says.

Now, he says, it's like there is a media woke quota. Every progressive channel is loading youth perspective and voice, angling for relevance with that growing demographic of people who are following and clicking and sharing. But it's still somehow striking the wrong note for him. Whether it's one-minute videos where people talk about social change in "a simplistic, idealistic way" or the Greta note that will soon emerge. We all know this note by now. It's the one in which the history of the youth climate movement has been scrubbed clean, where all active youth were inspired by Greta, replaced by Greta.

"There's been a whole, very revealing truth of how the media really selects their figures that they elevate and they put on platforms and, you know, yeah. Much respect and love for everything that Greta has done," Xiuhtezcatl says. "But it's much easier for the media to highlight a young, privileged, white Swedish girl than it is to talk about the young water protectors that are fighting pipelines in South Dakota or the many, many young leaders of color that have been on the front lines of this work for generations who have inherited these struggles. We're still trying really hard to kind of push through the barriers that we have at every level of the work that we do. From the political climate to the lack of representation in the media."

He is in Southern California recording his next album, so he is outside when we have this conversation on the phone. There is a backdrop of singing birds on his end of the call, and I can tell that he is walking, or at least pacing, because the reception has a kind of rhythmic dip. Xiuhtezcatl has just held two events with Greta. But what he is saying is that for all the humility Greta has in the face of her newfound fame, and for all the respect that he has for her, the iconography of Greta has taught him that he is never going to be as sympathetic as a white girl. For as far as the press could come in its coverage of the climate crisis, and of this movement, it has not come farther than lionizing a person who has the right looks to take the youth climate conversation mainstream. It erases him and a lot of other young people like him, whose American story perhaps has something more critical to say.

This possibility of erasure is an ominous prospect for *Juliana*, too, a profoundly important court case that has yet to see trial.

News coverage of *Juliana* has charted a similar path, as did coverage of the state-level cases in which Xiuhtezcatl was also a plaintiff. In 2011, what media attention lawsuits garnered was mostly about the novelty of the legal strategy, stories about audacity, about wackiness. By 2015, when the federal *Juliana* case was filed, the narrative about the wacky legal claims shifted to something like the backstory of a publicity stunt. In 2016, after the election of Trump, the *Juliana* story was clickbait. The administration's efforts to roll back environmental protections only fanned that fire. At its corporatized, underfunded, institutional heart, the news industry could not resist the lure of kids vs. Trump, even if they had filed their case against Obama.

Xiuhtezcatl doesn't blame any person when he talks about these kinds of failures. It's hard to imagine what he sounds like angry anyway. For as much as he likes to call out politicians, media trends, the deep-pocketed philanthropists who are fueling the rise of young white activism while inequities magnify in the Indigenous community, he is always going to be that guy who, two minutes later, can tell you to have a beautiful day and sound as sincere as when he was critiquing the media's woke quota. He is strident but unflappable, eloquent but not enraged. Sometimes I think that if everyone learned to speak like him, in such a smooth cadence but with conviction, probably we could all get a lot more done. But the kinds of issues he speaks to? Of debate vs. fact, of fad and fashion, of the fickleness of the donor dollar? He thinks they are rarely the fault of the people on the ground.

"When we come up with barriers and challenges within an industry as a whole, or the media, the news, it's not always going to be on the journalists that things are so broken and messed up. We're operating on a very corrupt system as a whole," he says. Image maintenance is, after all, something corporations spend millions on. "Everything from advertisements to making sure that the right story is depicted of them so that there's still enough misinformation, miseducation, lack of awareness from the greater public so that operations can continue as usual and they can continue to exploit as many people and as much of our communities' natural resources as possible." And indeed, social scientist Robert Brulle uncovered $558 million in direct foundation grants to climate disinformation efforts between 2003 and 2010, about three-quarters of it untraceable dark money. He also found that spending on lobbying by the fossil fuel industry topped $2 billion between 2000 and 2016—in other words, the first sixteen years of Xiuhtezcatl's life, most of them while he was deeply involved in climate activism.

Thus Xiuhtezcatl has seen this world of disinformation and fact bending up close and from a young age. By the time he was fourteen, he was getting threats from the oil and gas industry and there were months when his mother wouldn't let him walk home from school alone because she was afraid someone would hurt him. These conditions—our society with its levers pulled—haven't changed much. Not yet. As his work on coal ash and renewable energy gave way to fighting the fracking industry, to pushing ballot initiatives to increase the distance between fracking wells and homes and rivers and upholding the right of communities to vote for moratoriums on fracking, Xiuhtezcatl saw that Colorado's governor and various other office holders were also in the way. "We were up against everybody. Not just Republicans and Fox News. It was our own local media outlets who all of a sudden were saying they can't comment on this shit because it's too polarized."

Isn't this what's wrong with the articles about Greta, after all? As Xiuhtezcatl points out, they're not talking about corporate corruption, about money in politics, the foundational roots of capitalism. In other words, they're not talking about the core of what we're experiencing as Americans. Their authors want to do good. They want to highlight popular stories, not be alarmist. But they are frequently unable to really educate people about what is most wrong: that industry is deeply entangled with the leadership of our country and that this union is catalyzing the climate breakdown that could spell the end of us all.

In the history of the media's coverage of this issue, the diehards who have done this work are themselves marginalized and too few.

Now that we can see the effects of climate change—see the seas rising, the floods and storms and fires worsening, observe the melting polar ice caps and nations vying for the next best routes to ship cargo through places that were once ice—much public discussion about climate change has been cordoned off at its own water cooler, carried out by media outlets dedicated to the topic, like Greenwire. Punditry has risen with the obviousness of worsening weather, and it has filled the space where meaningful discussion once lay. One need only decamp to their favorite media venue for the confirmation bias they seek.

Meanwhile, CNN, MSNBC, Fox, all these networks feature theoretical smart guys and gals who are really in the employ of the fossil fuel industry, there to shill a message that's concealed by a few layers of think tanks and fancy degrees. The *Wall Street Journal, USA Today,* the Murdock publications, all have featured these types of voices. Even discerning publications publish unwitting industry op-eds when authors pose as ordinary people who also happen to be in the paid employ of the fossil fuel industry.

This has happened before in America. Makers of "trouble-free" asbestos, chemical companies, and flame retardant manufacturers all have a message to sell you. Now the drug companies. And, of course, tobacco.

Curious why people believe such things, or believe some things and not others, science historian Naomi Oreskes pulled ten years of science literature on global climate change to explore which parts of the climate "debate" were actually still in debate. She didn't find any. But when she published an article about the scientific consensus on climate change in 2004, she was attacked by people who called her a communist and clamored for her to be fired. Ever the researcher, she wanted to know who they were, so she researched some more. What she found were similar attacks on other scientists who had worked on acid rain, the hole in the ozone layer. And Oreskes also found the attackers were often the same. She coauthored the book *Merchants of Doubt* with Erik M. Conway soon after, outlining the history of these characters and their industry hopping. The work was later featured in a film by the same name.

She told the filmmakers, "All of this is a political debate about the role of government," pointing to scientists Fred Singer and Fred Seitz, two of the people who led the propaganda charge after leaving positions of prominence in the scientific community to work for Big Tobacco—Philip Morris and R.J. Reynolds. They were no longer scientists at all, but ideologues. And worse, people who'd spent their working years making weapons during the Cold War, who felt the threat of communism so strongly that they spent their later years attacking issues that called for government action.

Which is how scientists came to be saying that climate change was good for the planet between clips of tobacco ads claiming "Luckies make you healthy!" This in movie theaters across America, once *Merchants of Doubt* became a film. But it's important to remember that these were not climate scientists. They were scientists who held themselves out as experts in a field they had never worked in.

They and their cohort would surface again and again, on the eve of climate action, to insert their artificial "debate." In 1995, Oreskes found, just before the IPCC released its second climate assessment, scientist Ben Santer, who authored a key chapter, was attacked by such propagandists. In 1997, just before Kyoto, an absurd document called the Oregon Petition surfaced, purported to have been signed by more than thirty thousand scientists who did not believe in climate change when, in fact, no one knew who many of the signers actually were and some clearly were not scientists, claiming to be, for example, the Spice Girls and Michael J. Fox. In 2009, when Copenhagen was nearing an agreement, *Climategate* became a term many people knew but did not understand. It was not real, Oreskes told the cameras. Point being: any delay of action on climate change is, in essence, a derailment, particularly where international cooperation had been building. In other words, success for goons.

The press legitimized these fakes and pundits. James Hoggan, who would later write the book *Climate Cover-Up*, told interviewer Silver Donald Cameron on the Canadian show *The Green Interview* that few people ever see the social science on behalf of oil and coal companies. Or understand the massive public relations budgets that go into crafting terms like "clean coal" and getting people to use these phrases until even Obama is using them. It is a trick of heartstrings. The same PR tactics that are used to get people to buy things they don't need. They are tricks intended to trigger emotion and impulse. In other words, the triggers that usually drive people.

Hoggan founded DeSmog, a website "clearing the PR pollution that clouds climate science," which today runs the DeSmog Climate Disinformation Research Database, listing and profiling the individuals and organizations whose job it is to impugn the credentials, morality, or public image of credible climate scientists and their work. Among the 668 individuals and organizations in the database are names many of us now know. Like Steve Bannon, the former executive chair of that masquerade of a news service Breitbart, who briefly served as a strategist and counselor to Trump. And dozens of organizations like the Heartland Institute, the Cato Institute, the Reason Foundation, all funded by the fossil fuel industry and its wealthy donors, like the Koch brothers. All have been researched and verified to have been misleading the public and stalling action on global warming.

Google them, and you will find their op-eds in many a mainstream media outlet. Propped up, platformed, ready to inform you of whatever nonsense you have been missing.

This lack of institutional oversight has consequences. It's not hard to see that climate change has all the makings of the next major public health crisis. But what's worth remembering is that it also has all the hallmarks of previous public health crises. Like tobacco, climate change is an obvious public health threat that's being buried by a behemoth industry that has long spent its way out of problems. And like tobacco, that burial is attended by politicians entrenched in corporate-to-Capitol culture, who conceal science, issue poker-faced denials, and threaten litigation or other harms to hamstring critics and the press.

Many of us in this nation already know where this strategy leads. My family, like a lot of American families, had a front row seat.

I was fourteen when C. Everett Koop, the surgeon general who aggressively and somewhat roguishly advised Americans to quit smoking, was a regular on television, having just released his report on nicotine addiction. This means I was a teenager when my father screamed somewhat regularly at the television. He would be diagnosed with terminal lung cancer within fifteen years. Not C. Everett Koop, of course, who died at ninety-six because he had the good sense to follow his own advice. But my father, who surprised everyone, even himself, by living for twelve years with small-cell lung cancer, surviving on what seemed, at times, to be pure irascibility.

During this decade-plus reboot, there was acknowledgment that C. Everett Koop had been right about smoking. But in the early days of Koop's tenure, when Koop would appear on the nightly news with his roughly annual reports about how smoking was bad for you, there were times when my father would leap from his recliner and shout, over the length of his own extended arm, about the degree of misinformation C. Everett Koop was putting out about smoking. Koop was a renegade in this way. A Reagan-era appointee who popularized his role as surgeon general by upending the status quo on tobacco.

Though he'd been expected to toe the line, be a good boy about our good ol' boys in the cancer-stick industry, he issued eight reports during his tenure, many of them bombshells that included such details as how nicotine is as addictive as cocaine or heroin and that secondhand smoke is bad for you too. Thanks to Koop, smoking declined by 29 percent during his tenure, and the norm of the smoking section became as socially ridiculous as it had always seemed. These were the days when we had *smoking sections on airplanes*. And yet in a few short years, restaurants would begin banning smoking, and whole towns would soon follow, with some places even halting smoking within a few feet of the door to a building.

None of this was very popular at the time. The first of the restaurants to whisk up the ashtrays were boycotted and groused about by the cigarette-smoking patrons who decamped to the nearest places that still had ashtrays, then got nostalgic about how they liked it better there anyway and didn't know why they'd ever left. There were letters to the editor and a brief but absurdly misguided outcry for smokers' rights. And then we changed. All of us. People who intended to breathe healthily ruled the day in the end, while the smokers went home to their recliners to yell at their televisions.

Throughout this time, Koop was as hated as a public official has ever been. Yet he wrote the playbook on delivering disappointing news to an unwelcoming public. In other words, on how to uphold public health for people who were being endangered by the insidious marriage of corporate greed and political cowardice through no fault of their own. If he'd worked for Trump today, he would have been fired—or strong-armed into resignation, whichever came first. Controversy serves at the behest of the Commander in Brief now, who can dispense with appointees in fewer than 280 characters. Like poor old Rex Tillerson, the former secretary of state, fired on Twitter. But who can feel bad for Tillerson, former Exxon CEO, who perhaps lucked out in avoiding the charge of a State Department soon to revolt. Intelligence chief Dan Coats was perhaps more sympathetic. Or maybe acting secretary of defense Patrick Shanahan. Or White House counsel Donald McGahn. All left the administration via Twitter. Maybe they saw it coming. Maybe those really were announcements . . .

Meanwhile, you might wonder, as I do, where *is* the US surgeon general on climate change, anyway? Some are present and accounted for. Like Surgeon General Vivek Murthy, who tried to rally the public around the threat of climate change in 2015 and 2016 under Obama. And the two surgeons general—David Satcher and Richard Carmona, appointed by Clinton and Bush respectively—who gallantly coauthored a pro-*Juliana* op-ed in the *New York Times* and later an amicus brief.

But they don't have the same effect because no one is surprised by what they say. All were appointees of the last presidents to believe in climate change or their reports were tamped down—as Carmona claims his were—and scuttled. Thus they are not renegades or boat rockers, whistleblowers or cage rattlers. They are rank-and-file types, doing what's expected or trying, anyway. Koop's special sauce was more that he was a lot like Trump. He was a master of surprise, of audacity and gall. It just so happened that he had a purpose beyond only that.

Which brings us to the docile Jerome Adams, appointed to the Trump administration in 2017, a pick by Vice President Mike Pence and a well-mannered Indiana sort who has also not rocked this boat. He has done good things: made the opioid epidemic his primary duty, for example, along with mental illness. But he hasn't rallied the nation around Earth's warming atmosphere, the melting ice that is making the seas rise, the floods and storms that are only getting worse, or the forests that might contain this damage but instead are going up in flames. In my world, these are the things that make it increasingly difficult to stay off opioids and keep my mental health together, no kidding. But maybe that's just me.

Good thing for Xiuhtezcatl, to remind me that I have spent too much time listening to the wrong conversation. Because the story of the young, despite the harms they face, is always a story of hope.

"We can't be that dire," he says. "I think the fact that young people have been left out of this conversation for so long, that's why there's still these flaws in how the media is telling stories around the climate crisis and the climate movement." If they listened to the young, he says, they'd hear clarity about solutions, hear optimism, and leave behind the hopelessness and the despair that are trademarks of this conversation's past.

Young people, Xiuhtezcatl says, "cannot afford to lose hope and to be apathetic and disconnected from this issue, because it is so urgent for us." Not in a hundred years. In the next five. In California in the fires. In the Gulf experiencing the direct effects of flooding and worsening storms. In northern communities under thaw.

He believes the fight needs everyone, every skill, and that even if not everyone can be an activist, every person can show up in simple ways. For him, he has immersed himself in hip-hop as a way of crossing into communities that would not otherwise have this conversation. He raps in "Boombox Warfare" about the legacy of resistance, the speed of the planet's warming an analogy for his own rise. This to the unlikely accompaniment of marimbas, of violins.

"If you look at history, artists have pioneered and been the voices of these movements," he says. Bob Marley. Tupac Shakur. John Lennon. If the fight needs musicians, too, Xiuhtezcatl has decided that will be him. And the other place he wants to fight is in the courts.

"Let them take their time," he says. "I have a relatively interesting amount of optimism when it comes to the courts acting on behalf of our climate and our future, just because of the rulings we've had in the past and how much we've just defied people's belief in what was possible."

Props to him. And for all the optimism of the young. After climate change begins trending, it changes public opinion for good. In 2016, only 47 percent of Americans believed in climate change. By June 2019, 70 percent want the United States to take aggressive action, with a majority calling for a full transition to clean energy within the coming decade.

Instead of the language I learned, a medical dialect of PET scans and percentages, of bronchoscopies and chemotherapy, pulmonology, nephrectomy, liver and bladder resections, they have another tongue. A call to action. A language of chance, of hope and opportunity:

#saveourplanet
#climatechangeisreal
#climatecrisis
#climateemergency
#extinctionrebellion

CHAPTER ELEVEN

The Zenith of Ridiculous

A month after the canceled trial, there is still no word from the Ninth Circuit about whether it will let the *Juliana* trial go forward. As the waiting continues, wildfire is still burning in California, killing scores and decimating the town of Paradise. And because disaster is the low-hanging fruit of the news industry, and I am unemployed now that the trial is canceled, I head south for the work of covering the aftermath for Reuters.

In homes reduced to ash, tree trunks singed black, grasslands wiped out by flame, I will get a look at the apocalyptic world Isaac foretold. I will also come face-to-face with Ryan Zinke and Sonny Perdue, secretaries of the interior and of agriculture, respectively, and the only defendants in the *Juliana* case I will ever stand in the same room with. I will be able to ask Zinke, in person, if in the face of such devastation, America can change. It will be a profound moment, one that alters my perception of the climate fight for good.

Though climate breakdown created the conditions for this fire, mostly through extended heat and drought, the discussion about its role was reduced to political joust in the first days of the burning. On Twitter, Trump blamed poor forest management—ignorant, it would seem, that the flames were mostly carried by wind and by grass understory, before burning house to house. On social media, red-state friends picked up the thread of "forest mismanagement" and began jabbing it like a poker off hot coals.

Days later, when Trump stood with California governor Jerry Brown and governor-elect Gavin Newsom at a press briefing and made absurd, indefensible remarks about how the president of Finland had told him that nation prevents forest fires by raking the forest floor, he made a smooth, soothing side-to-side motion with his hand, as though combing dirt is really all it takes. Beside him, Newsom suppressed a reaction that first looked like laughter but then more like a stifled scream.

What followed was a week of ironies. On Monday, the IPCC released its report cautioning that humans have only twelve years to take the steps to halt climate change, with UN scientists sternly warning that without urgent, drastic action, the world would see mass die-offs of coral, insects, fish, and plants, a combination of events that would spell food shortages, disease, and worsening disasters as sea level rise threatened ten million people. The next day, the Trump administration announced plans to expand logging and thinning on federal land, despite countervailing science showing trees would be critical to sequestering enough carbon to avert climate breakdown.

Then on Thursday, for Thanksgiving, volunteers made fifteen thousand Thanksgiving dinners for people displaced by the Camp Fire. On Black Friday, a strategic date, the federal government quietly released—aka buried—its Fourth National Climate Assessment, acknowledging the increasingly worrisome realities of climate change. And that day, I drove to Paradise to see for myself how the West will look when it is ravaged by flame.

On the drive south from Oregon, NPR reports that the burn area around the Camp Fire is bigger than Chicago. Rain is hitting California as a fresh relief and by the time I arrive the fire is out. Once I reach Chico, it takes me two full days to get a grip on the size of this disaster, which is my job, that and reporting on the recovery. It isn't easy.

There are thirty-four thousand evacuees—those from Paradise, and Magalia to the north, and other mountain dwellers who were outside city limits. Now, the ones who haven't left the region have combined with rescuers to overrun every hotel within roughly 30 miles of Chico, plus most of the campgrounds. Every apartment or room for rent that can be found in a county with a 2-percent vacancy rate for rentals has been had. Evacuees have also dried up the market for houses for sale, and trailers and RVs, too, buying whatever they can drive or tow and park, then live in.

The recovery effort is immense, dizzying in its scope. It includes the National Guard, the Red Cross, and FEMA, plus local agencies and charities too numerous to count. The fairgrounds that are used for staging operations and for shelters are spread over three towns, each a half-hour drive from the next, making even the geography of reporting on it challenging.

Command central for disaster response is set up in Chico at a fairground known as the Silver Dollar, where the marquee is still advertising whatever gun or fashion show was the most exciting thing happening here a few weeks ago. The parking lot is awash in official-looking vehicles. There are police cars and fire trucks that look sturdy enough to drive over the moon. Several rugged ambulances are equipped for search and rescue, belonging to the teams from five states that have come to help. There is every type of monstrous-looking machine that has utility in a situation where roads have been buried under abandoned cars, downed utility lines, trees, and rubble. These include vehicles from the National Guard, whose canvas and crescent-roofed pop-ups are assembled in a field outside the fairground's gates. From the street, it looks like a set from the film *Arrival*, a place where one makes contact with another world. And in a way it is.

In the final tally, 153,000 acres around Paradise have burned. The western edge of the scar traces the road that runs from Chico south to Oroville, speed limit 60 miles an hour. Still, it takes seven minutes to drive the border of the burn at that speed—it is that large. Its size is otherworldly, almost incomprehensible. And it extends east from the road as far as the eye can see. When I look at it for the first time, I nearly veer off the road just gawking. People stop, take pictures, try to take in the enormous dimensions of it. But it is impossible. The few strands of grassland and the deep tractor-tire marks along the edge are all there is to indicate that this is no freshly plowed field, that what's happened here is a blaze barely contained. In a few places, the fire jumped the highway for a few acres to the west and nearly made a getaway toward more towns. Amazingly, the trees still stand on the ridges that roll toward Paradise.

For now, the town is officially closed, with 18,744 structures burned, 13,965 of them people's homes; search and rescue teams are looking for hundreds of people still missing. Twice a week, Butte County officials hold press conferences in the main hall at the fairground, called Harvest Hall, a flat-looking building with a utilitarian floor that can be piled up with folding chairs and TV cameras and a podium to create some kind of order. By Thanksgiving weekend, it is clear that the people associated with this effort have worked nonstop and are exhausted. Like the press, they rearrange their shifts and duties after the holiday, sending in the reserves, so many people arriving turn out to be people like me—people who have just begun to wander these encampments of crews in imposing uniforms, safety orange and hard hats, trying to get a grip on whatever it is they are supposed to be doing.

Officials from Butte County have mostly taken charge of the efforts under way: looking for bodies, figuring out when and where school will resume, taking stock of what buildings still stand, where people will live, cleaning up the debris, and going about the business of restoring utilities—power, phones, water, natural gas. It falls to the county sheriff to oversee the search and rescue, and to find all of the people who are unaccounted for. Just days ago, that number was topping twelve hundred, but every day it falls—through a combination of recovered bones and a desk-and-shoe-leather effort that finds people who are staying with relatives, in campgrounds, or elsewhere and didn't yet know they had been reported missing. To grasp the number of people still not found is staggering, stultifying. Unsettling. And day by day, the sheriff's reports to the press are followed by a parade of other experts, in charge of all the other parts of recovery, reporting on their own next steps. Some are shelter officials, offering head counts and details on which shelters remain open and where, weathering their own outsized efforts.

Not everyone is comfortable with this officialdom. Just down the road is a Walmart parking lot where people resistant to shelters have set up a camp. Some are camping in cars and others in tents and tarps that first converted a trapezoid of grass between the store and the highway and has since slowly swelled into a tent city for refugees. By now,

Black Friday is over and there are security vehicles, SUVs driven by official-looking sorts whose job, it seems, is to cast cool but stinky eyes in the direction of the campers who remain while alternatives are on offer. Norovirus has hit the shelters as hard as it hit Paradise before the fire, and some of the campers say they would rather take their chances with the rain than the germs. Plus, the location of the Walmart is hard to beat.

The Walmart, like a lot of Walmarts, is a barge in a sea of pavement, surrounded by other barges. Lowe's and Kohl's and the Chico Mall, the parking lots of which also fill up with disaster aftermath. Insurance companies set up mobile claim centers—MCCs for short—in RVs, trailers, even air tents, in the parking lots of the big box stores, serving hot beverages and cookies to displaced people who turn up to begin dealing with insurance, if they have it. Evangelicals—initially with help from the secular set—have taken over a vacant Toys "R" Us building and are using it as a clearinghouse for supplies. Camping gear and bottled water. Religion-themed disaster kits and clothes. Food, blankets. And gift cards—lots of gift cards. They are calling it Miracle City, but the guy who runs it says it's more like running a grocery store in hell. The staff wear red armbands to distinguish themselves from the survivors, whose arms are taped with blue. As I talk with the boss about operations here, a staffer passes carrying a cloth-wrapped container filled with gift cards to the safety of the primary-colored checkout counters. A man in a neck brace has been given a table at which to examine people's eyes, then fit them with replacement glasses.

FEMA is across the road, set up in the old Sears anchor store on the east end of the Chico Mall, which has quickly filled with one-stop shopping for the disaster ravaged. Besides running its own programs, offering shelters and other housing aid, FEMA offers space to donors like St. Vincent de Paul and the Red Cross, along with a lot of other entities. They offer medical equipment, therapy dogs, directed play for traumatized kids, and various loan programs designed to get people and businesses back on their feet. This is where the sheriff's office is collecting DNA from the relatives of the people still missing, and where the state's vital records department is reissuing things like birth certificates to those people who no longer have them. The DMV is there, too, to make licenses from the vital records. And when lines get long, the place works just like a DMV: people take numbers and sit in neat rows wearing glazed expressions. Many carry dogs and suitcases, random donations and water. There is blue tape on the floor to route the ones who can no longer pick up their heads.

To be clear, I have never seen anything like this. And after two decades in journalism, I didn't think there was much I hadn't seen. My assignments have included all manner of death and destruction: landslides and floods, active shooter incidents (three of them), and the kinds of random accidents and crimes that befall people who don't expect these things to happen to them. Disaster on this scale, however, is something new. To say that any of the rest has prepared me for it would be a lie, though I thought it would have. As I write stories of escape, of Pepsi trucks commandeered by police and parents driving children helplessly through backcountry roads, chased by flame, I find myself, in moments alone in the car, blasting some of the worst pop music of our time just to convince myself that I am in a better mood than I actually am. Right now the days are just adrenaline—too much to do and not enough of me. But later, after the job is over, I will notice new reflexes: eyes darting to dark patches in grass; gaze lingering on parking lots, the neat rows of cars still wearing their paint; inexplicable weeping at unscarred trees. But for now there is just the work.

When I stop at a campground in Gridley, which FEMA has converted to a shelter, a woman with long white hair approaches me from behind, carrying what looks like charcoal briquettes, and remarks that at least the ground here is flat. She says she's been to two other campgrounds so far and both had hills. Then she says: "I've been burned out of my house. It was scary. There was a big fireball coming for us," and keeps walking.

It is like this everywhere. It's been two weeks, but even simple questions about daily routine, about services and recovery, quickly drift backward, like wormholes through time to the day of the fire. Simple conversations about insurance, mail—the practicalities of uprooted lives—pivot on the words "I was just . . . " or "I was at . . . ," and what follows are tales of escape, of near death. They are the detached narratives of people still coping, still making sense of what is behind them.

Soon, a language develops. In a Safeway restroom, a woman tells me, "I'm burned out." It does not mean what it used to mean. She follows by explaining where she is living and how far that is from where she used to live. I wish her luck. And in this way, the pleasantries of disaster take form. They are more than polite exchanges. They are a way to communicate one's relationship to collective tragedy, a form of navigation. A kind of Marco Polo. *Is anyone here with me?*

The Chico post office becomes a kind of waiting room, a place to anticipate contact from the outside world. I stand in line and talk with people as they snake their way down the sidewalk, up the stairs, and into the building to collect their mail. Some are still living in shelters, still trying to find housing; others are attempting to understand next steps from the insurance companies who have yet to be let into town to survey the damage and issue claim checks.

"I kind of feel like a turtle on its back, looking for a rock," says Ed Riddle, fifty-three, who had the loss of his home confirmed by drone images and by a friend who works for PG&E.

Christopher Gregg, forty-two, is living with his wife and two children in an RV park that placed them adjacent to the facility's burn pile, a choice he resents. He hasn't had time to mourn the loss of his home or his neighbors, whose bodies had been discovered in theirs. He says he misses walking to his mailbox.

Thirteen-year-old twins Makenzie and Bradlee Shaw and their mother, Roberta Talley, are searching animal shelters for one of their three dogs. Their house is gone, incinerated, they say. Their other two dogs were in their kennels when it burned. When their mother shares this detail, both twins look away.

This is the context in which, at the federal press event, I find it hard to behave myself. I am used to this: to scripted conversations masquerading as discussion, the dog-and-pony and the pomp-and-hold-your-questions. But it is Monday and it is daylight, finally. The rain is not falling and the smoke has cleared, the fire gone, and today is one of the first days to take stock of the state of the town in the light and clear air, while reporters are among the few still allowed in.

When I drive into Paradise, climbing the road from Chico for the first time, much of it is still covered in ash, though its tree canopy still stands, the bases of the trees singed by flame. The roads have been cleared of wrecked cars, but many still line the shoulders. The paint and tires have been burned off so that they are a kind of uniform color—ruined metal—and the aluminum rims are melted, warped. It is hard to describe the ash. It is less like ash as I know it—a dusting of dried embers—and more like filthy snowdrifts. It piles around the still-standing chimneys, forms a thick understory to the twisted metal of onetime roofs and footings, or just strikes the shape of the thing it used to be: ash in shape of house, ash in shape of garage.

Without a map by which to follow the signs, I find it hard to stay oriented, to maintain a sense of direction. I wonder how long it will take for anyone who knew this place to replace their visual cues—gas stations and restaurants, parks and homes—with the metal fragments and trees that are left. This is why I am here: to bear witness, to see. And instead, the Department of Agriculture closed the road today and I could not enter the town unless I RSVPed for the government stump speech and held my questions.

I go to the Paradise Alliance Church, which is still intact and where the required event is held, and watch the SUVs ferrying Zinke and Perdue to the show. These guests are important enough that there is a staff huddle outside the church, and a serious but stern countdown of the minutes until they arrive. Almost everyone is wearing woodsy, outdoorsy outfits—staff and appointees alike—though the layers of makeup and the boots that have never been dirty make it easy to tell the policy wonks from the locals. I run into an old colleague, sit beside him, and inquire about life since I saw him last. Then the inevitable "How long have you been here?"—which is code for the thing we do not ask: *Are you wrecked yet?* He tells me he is exhausted after eight days and that this is his last assignment before he can go. He says he found bones, that he just has to make it through this last briefing. And I say it is a good thing to find bones, to account for a person, but he gives me a look that tells me I am making things worse.

Then the show begins. I do as journalists do and quickly sort the camps. The federal camp. The state camp. The locals. I note the four placards quickly removed from the seats reserved for county officials—the people largely managing the disaster—who likely have too much to do to show up, my first clue that what I'm about to witness is a bunch of nonsense.

In short, the exchange goes like this: the feds want to thin more timber and the state reps say they have already done so and the locals are just trying to be real without getting crosswise with the rest because the rest control the money for the recovery. I can't imagine how the mayor, Jody Jones, is doing this, sitting in a church in her decimated town and wearing a poker face while everyone else pretends to know what is good for the place. Still, there are hotshots in the room, the elite wildland firefighters, and they don't give a fuck. I love this about firefighters, about police too—they have no filter, no talking points; they will tell you what is real.

Today, they say: "Forests are thick, but when it's hot, windy, and dry, fire burns." This is Leland Ratliff, thirty-six, who says the hotshots are understaffed to do the work that the federal government is prescribing. He's had a thousand hours of overtime in six months just fighting the fires, and the deployments are long and hard. "Our families struggle and we struggle," he says. Fire outside communities? He says we should let it burn.

"This fire, if it happens again, we're not going to stop it," Ken Pimlott, Cal Fire director, assures. He is the top firefighter for the state of California, a state that fights wildfire annually, and arguably twice a year now, in summer and fall. "When fire is burning like a blowtorch, our traditional tools are not going to work."

And this: "Wildfires are not just forest and wildland problems anymore. They've become city problems," says Jim Nelson, a state senator, in a nod to the effects a warmer, drier climate has added to the wildland interface, something experts will echo later. "Urban areas are suffering what we in rural areas have long suffered."

This is supposed to be a community conversation, but the community isn't here. The community isn't even allowed here while the town is closed for safety. Really, it is just the press in a room, TV cameras looming over the heads of the print and radio people, while the various government types pretend to be saying more than they are really saying. After an hour, when the hotshots have said their piece, the one black man on the panel has been overlooked and becomes the only person to make no comment, and the federal appointees have dutifully imparted whatever messaging has been crafted by the people who wrote the messages, someone announces that the meeting is over and there won't be any questions until after a tour that is, in effect, mandatory for those of us with questions.

All I can think is: *What the ever-living fuck?* Because we've been one hour in a room without anybody saying anything, the light is fading, we have work to do outside, and now they are telling us that protocol puts us an hour from any meaningful back-and-forth.

I am torn. Everything in my training tells me to behave as they wish. I have been to many a press briefing, endured many a nonsensical speech, and been bludgeoned with procedural inefficiency, exhausting timelines, and imaginary constraints ("We're so sorry, but his plane leaves in forty minutes") for long enough to know that this is just how it is done. Political pageantry is part of the daily practice of keeping the press at arm's length, all while shilling one's message. And always, eventually, I find my way in. But this particular situation is galling. At last count, eighty-seven people are dead and more than four hundred are still missing. While we wasted the last hour, search crews were combing the ash of more than eighteen thousand structures in search of remains—*eighteen thousand*—all while federal officials are stumping for plans to expand logging on federal land, something the head of Cal Fire, Ken Pimlott, who is four seats from Zinke, has just told us is funded, has already been done, and would not have had an effect on this situation.

There are more important things to write here: about how much longer until everyone is accounted for; how long before people can see where their homes used to be, until insurance companies can get drones in the air and start settling claims so that people can move on; and how the five thousand schoolkids who've been displaced from the thirteen schools that burned, or partly burned, will be educated, and when and where.

The list goes on. And when I look around at the other members of the press, I think we are not behaving badly enough. Scuttling to our cars, instead, on some tour of a solution to a problem no one has. This was not a typical forest fire. It was a fire on the fringe between a town and forestland, one fueled by homes and backyard tinder as much as natural fuel. This is an entirely new problem, a growing one as humans build deeper into woodland and towns run into towns. And it is exacerbated by climate change. Worsening heat and drought make these spaces more dangerous than in the past, and we've yet to design exit routes or set standards for construction—like metal roofs, for example—that can mitigate these dangers. A few days ago, this very government released a report acknowledging these realities. And here we are, amid a climate catastrophe, politely allowing them to deny it.

I ambush Zinke. I have never done this before, broken protocol with a federal official. But I do it now. Maybe I should have done it a long time ago. As he is leaving, drifting out from behind the swath of tables and making polite banter with the people who are allowed to speak to him, I pick up his tail. I follow him out the back door, the one the press is not supposed to use, and onto a narrow staircase where his staff accidentally lets me pass and where he is now stuck with me. I ask him, "Secretary, do you have time for a quick question?" And he is a politician, so he says, "Sure."

Now we are at the bottom of the stairs. He turns to face me outside the SUV he is being ferried in. And this is the exchange we have:

ME: "You mentioned lengthening seasons, increasing temperatures . . . "

ZINKE, interrupting: "The fire seasons are longer than they used to be. The temperatures are elevated."

ME: "Sure, yeah. Moisture content's lower. Beetle kill . . . "

ZINKE, interrupting: "And that's a background to, then you have tree density and the density of trees."

ME: "Sure. But is that a nod to the climate change report that came out Friday?"

Here the woman with the nice shoes tries to interrupt us. Her name is Faith, which strikes me as ironic. She has an iPhone and has turned on the recorder so she can catch what is coming out of Zinke's mouth, presumably to correct me if I mishandle it later. But there's no mishandling this. He is mishandling it just fine without me.

ZINKE: "I think you have to take everything into consideration based on science. There's no doubt that temperatures are elevated, we're in a drought period. So, to me, this is really not about climate change. This is a backdrop of something that could or could not be under our control. It's indisputable that temperatures are higher, seasons are longer, the moisture content is there. That's just basic science and metrics. Now given that, what do we do about it?"

Later I will Google this deflection. And I will find similar words uttered by Sheryl Corrigan, who holds the Orwellian title of Director of Environment, Health and Safety at Koch Industries. She spoke about Charles Koch's climate beliefs at an event hosted by the *Wall Street Journal* in April 2016. And her conveyance of those beliefs sounds eerily dutiful, like a cult member parroting the language of a fringe, unseen leader. Here's what she said: "Charles has said the climate is changing. So, the climate is changing.... I think he's also said, and we believe, that humans have a part in that. I think what the real question is... what are we going to do about it?"

I won't know this until the day I think to Google Zinke's comment. For now, I am just standing at the bottom of the stairs while Faith tries to shoo me away but I am close enough to really take him in. I learn that besides being good looking and affable, Zinke has another quality that winning politicians tend to have: a spotless, affected aura. In a politician's arsenal of superpowers, it is one of the three that really matter. There's the power to deliver a speech, the power to transform oneself in front of a camera, and then this: the power to stand next to people. So even though I have never met Zinke before and have ambushed him on the stairs, I instantly have the sense that we are friends, that I have known him forever, and that maybe he isn't such a bad guy, even though he is the subject of a disturbingly high number of ethics investigations and has just gone off on a tear about how logging and cattle grazing reduce fires. He's like a drug. So while he's talking, I nod at him, even while what he's saying makes no sense for the situation we are in. You can't graze cattle in a woodland town, and the Camp Fire burned house to house in Paradise, not tree to tree, something evident from the fact that there are trees still standing all around us this very moment while the houses are ash.

Still, I have got him off script and it is all I need. Whether I quote him or I don't, all I want is a moment to understand whether this crap that comes out of his mouth is something he believes or something he is required to say. And I think I see, in these few moments of unscripted babble, that Ryan Zinke knows we have arrived at the zenith of ridiculous. And that what is passing for leadership these days in America is just political theater, a carnival of convenience run by people whose agenda will be unchanged by this calamity.

But I can't know that, really. And next they are whisking him away.

Later, after I skip much of the tour to file Zinke's comments and see, finally, the ruin of the town, I catch him at the Q&A. There, my colleagues push him hard on climate change and whether it is related to this fire. By then, Zinke has shored up his talking points and repeats his earlier comments, somewhat more eloquently, saying he believes in science and how there are conditions of rising temperatures and drought that are a backdrop to this fire, but whether climate change matters to it all is a "debate." But there is no debate. Just three days ago the Fourth National Climate Assessment, compiled by roughly three hundred experts and overseen by Trump's own administration, confirmed that climate change is real, its effects devastating, and those impacts, if not already upon us, are imminent. The only people still avoiding that truth here today are the two federal appointees at the podium.

No one uses the term *ecoterrorist*, but Doug LaMalfa, the Republican congressman representing California's northern interior, lets loose some sort of war cry about the environmentalists who are disrupting fire recovery by standing in the way of timber salvage.

By now, we are standing at an intersection near what used to be a glass shop. The glass is melted in sheets in the rubble. It is art glass, the kind of glass people shape with a blowtorch, and I know this only because a friend who sells blown glass told me that a client lost a forty-year-old shop here a few days earlier. To see the blue plates standing like stalagmites on the concrete, steel girders in a knot, drives home exactly how hot it was here while the fire burned. Blowtorch hot. Aluminum-melting hot. Twelve-hundred-degree hot. Nearby PG&E is reposting the telephone poles, stringing new electrical wire where the old wire hangs in the trees. Meanwhile, the most damaged of these have been marked for clearing with orange Xs. Cleanup crews have been clearing trees for days now, and they are otherwise discernible targets, obvious even to me: their bark is charred and flaky like mica and they are clearly too unstable to be left standing. No environmentalists are in the way. LaMalfa ignores a follow-up question about whether these standoffs over salvage timber have actually occurred. And again, there is no more time for questions.

The mayor is next to me and turns away when it is over. She has incredible poise but when I am close to her, I can see that she is exhausted, steeling herself. She's been giving her community a thing it needs lately, something that must be unfathomably hard to provide: assurance that someone is in charge, plus certain words about how Paradise will survive it all, that home is still here. Responding to plays on "Paradise Lost," which the press and a lot of other people find impossible to resist now, she earlier listed all the things Paradise has not lost: its town hall, two of three fire stations, the library, two of three grocery stores, the post office, the banks. She even worked in a quip about the Starbucks and the Dutch Bros.

She has lost her home too. All of the town council did, along with 90 percent of the rest of Paradise's residents. I wonder then what thankless job Mayor Jody Jones has lucked into, and for how many years. I wonder if anyone pays her, if she will hate it, if she hates it already. Then I wonder what nobler or more important thing she could do, maybe ever.

It strikes me then that leadership is often foisted on people by circumstance, not a thing achieved. That true leaders are people like her, people who step to need. And that there is no one here from the federal government who fits that description.

CHAPTER TWELVE

A Dislocation of the Soul

Miko Vergun has three words for all of this: "Solutions! Solutions! Solutions!" A *Juliana* plaintiff from Oregon, she says she doesn't care whose fault it is. Or how we got here or why we stay the course. She only wants a fix.

She says this sitting on a couch in the visitor center at Tualatin Hills Nature Park outside Portland, her arms pulled into her sweatshirt and a fire in the fireplace in the tiny library there. Miko's not a loud kid. We've just taken a walk with her mother, Pam, and brother, Isaac, also a *Juliana* plaintiff, and Miko has been quiet for most of the morning, deferring to Isaac to fill in the gaps when she was tired of talking or just tired of trying to articulate what she wanted to say. But when I ask her the question I ask all the plaintiffs, the one about how they feel about past generations' role in the climate crisis, Miko gets feisty. This kind of navel-gazing—intellectual meditations on responsibility, on blame—is an obstacle to solving problems, she says.

She lives in Beaverton, a suburb just outside Portland, a place that—like Portland—is sometimes afflicted by a left-leaning righteousness that can tip into made-for-TV hysteria. Thus, Miko's world can be one of fanatical social causes, her irritation proof that even liberal places can become intolerable on the subject of Earth stewardship. She describes the jousts and one-upmanship about how to be your best environmental self. The debates about whether someone can, for example, be a climate activist and also a meat eater. And then she talks about how much this doesn't matter. How we need systemic change, not individual quests. And how exhausting all the infighting is. How off-key it sounds in contrast to the worst of climate impacts in the world. Then she sits back, exasperated, and says, "Great. You recycled. Want a cookie?"

It's hard to blame her for this angst. Miko is Marshallese, adopted into the United States at birth. At the behest of the documentary crew and on the eve of the would-be trial back in October, she traveled to the Marshall Islands with her mother to meet her birth family and to conduct her own interviews about how climate change is affecting the islands. It was her first time in the country, and on the trip she was faced with exactly the kind of lopsided impacts she describes to me now. (Vic Barrett, testifying to his own Indigenous roots in a remote part of the world, will soon describe these same things to Congress.)

Miko describes her birthplace to me. "It's during the spring when everyone crosses their fingers and hopes that everything's okay because that's when the king tides come in," she says. King tides are normal on the island. But for the Marshallese, "because they're so close to sea level and storms are stronger because of climate change, it comes in with more velocity and more water. So it's devastating for the people that live right next to the water." The island is pretty narrow, she says, so people living on the coast are more at risk. "And there's no mountains or anything, it's completely flat." No place to run to, no place to pack up and move.

The Marshall Islands—twenty-nine coral atolls in the Pacific, home to more than fifty-eight thousand people—are a mere 6 feet above sea level. In the last thirty years, sea level has already risen a foot around the islands because of changing trade winds. Climate change promises more as polar ice sheets melt, oceans warm, and the warmer water expands. Episodes of king tides have already swamped homes, downed seawalls, and washed away graves. In some areas, salt water is killing breadfruit trees, wiping out this crop that's sometimes used for subsistence, sometimes for cash. Drought is worsening in parts of the islands at the same time. And as storms grow more powerful, they compound damage to coral reefs and fisheries already affected by ocean acidification and rising temperatures.

The options ahead don't look good: move or build higher. Plans to elevate some portion of the islands have so far not found financial support from the developing world. In one projection by the US

Department of Defense, which has a military base in the Marshall Islands, at least one of the atolls will be unlivable by 2030 if the Antarctic ice sheets continue to melt. These are not distant threats. They are scenarios that have already arrived for the Marshallese, people who live day to day in tense negotiation with an encroaching sea.

Miko saw this firsthand when she went for a swim at the beach with her family, saw how her aunt was afraid of the water. How the tide crept over the sand, a strip of muddy shallows the width of a road covering where the beach used to be. A dozen paces into the water, the seafloor dropped precipitously, like the edge of a wall. It made the local people—people who knew how quickly a rogue wave could destroy a home, wash a grave to sea—nervous to stand close.

When Miko interviewed a youth group leader, "She was saying it's difficult to live . . . when you're living in fear of whether your house is going to go in the water."

These are the kinds of outsized problems Miko contrasts with the rank-and-file do-goodery of life in a progressive American suburb. And with debates about hamburgers. To her point, they are not just issues for other people, her people—things that happen on far-flung islands 2,000 miles from Hawaii. Though the most disastrous climate impacts do disproportionately befall people in other nations, on small islands, and in the Arctic and Africa, too, these harms affect Americans every day. And not just because the United States has a long history of immigration, is full of diverse citizens like Miko and other *Juliana* plaintiffs who feel the acute pain of climate impacts in the places their families come from. But also because climate change is spurring migration, not just *to* America but within America, in ways that aren't so easy to see.

One example is that a third of Marshallese people already live in America, spurred to move by economics and opportunity, chiefly to Guam and Hawaii, where they are welcomed by a US compact that offers free passage as compensation for missile testing on the Marshall Islands after World War II and in the sixties.

In Hawaii, scientists predict higher temperatures will spur heat-related illnesses like dengue fever and cholera and stress the plants and animals, making way for invasive species. A decrease in winds is expected to disrupt rainfall, causing droughts and periods of heavy flooding. Warmer oceans and ocean acidity could disrupt marine ecosystems, too, and thus the food supply. All that comes with the stark reality that rising seas will plunge beaches underwater in time, thus tanking an economy that mostly turns on tourism for residents like Journey Zephier, also a *Juliana* plaintiff, whose cultural practices are also threatened. Journey is a citizen of the Yankton Sioux Nation who has been living in the small village of Kapaʻa on the Hawaiian island of Kauai for the last decade, adopting its language and customs like fire dancing and Tahitian drumming. The island is another place with dying coral and shrinking beaches, set to be underwater by the end of the century.

In 2016, the White House Council on Environmental Quality co-hosted the Symposium on Climate Displacement, Migration, and Relocation to advise Hawaiians on how to "plan for and implement voluntary migration" and to address the legal and policy challenges for those for whom relocating will be the only choice.

Miko knows how heart-breaking these decisions can be. So from American mainlanders who live without such problems, she doesn't want the complaining, the intertribal politics, the partisan bickering, or the blame game about whose fault climate change is and who is doing the most or the least to tend their own garbage patch. She wants policy, results, action large enough to assuage not mainstream American guilt but real suffering.

This kind of displacement—the forced kind—is increasingly common as climate conditions worsen, and not just in far-off places like the Marshall Islands.

Of the 16.1 million people displaced by weather-related disasters around the world in 2018, 7.5 percent—or 1.2 million people—were Americans. Hurricane Michael displaced 464,000. The Woolsey Fire in California uprooted 182,000, and the Camp Fire in Paradise another 50,000. Over the years, FEMA has bought tens of thousands of people out of homes damaged by Hurricane Katrina and Superstorm Sandy. Looking ahead, scientists predict as many as 13 million people in the coastal US could be affected by sea level rise if the polar ice sheets collapse, possibly spurring "US population movements of a magnitude similar to the twentieth century Great Migration of southern African-Americans."

Citizens in the Navajo Nation are an example of US mainlanders who face these kinds of frontline risks. And those risks are a primary reason that Jaime Butler, who has lived much of her life thirty minutes north of the Grand Canyon in the Navajo Nation, is a *Juliana* plaintiff. Unlike the Marshallese, Navajo leaders have found a range of possibilities to adapt to climate change on the reservation. They're working diligently toward climate redress, and their planning and execution is illustrative of the kind of solutions other communities could undertake in the climate fight.

What such action aims to preserve is this: Jaime on her grandmother's land as a girl. The flat, dusty earth at her feet, blue sky overhead, and sheep grazing on what plants will grow, days a rhythm in and around the hogán. Jaime does what jobs need doing, her cousins too. She takes care of the sheep, cooks food, fixes things, feeds the herding dogs, and keeps a lookout for the coyotes and wolves whose job it is to try and get the sheep.

Her Grandma Eleanor speaks only Navajo, or mostly Navajo with a bit of English sass, so Jaime has to guess at what her grandmother is saying, intuit what it is she's supposed to do. This barrier between the languages makes it easier for Jaime to fall into the life of her people. Less distraction, no chatter. It's a path for Jaime to herd sheep all day and just *be* on the land in the way Diné were made to be.

"It wasn't like an actual camp, it was like our parents sending us to our grandma's house to help her. We called it sheep camp. I guess that's just the word," Jaime says. The Navajo are a matrilineal society, which means that women have traditionally owned property and sheep, and that a person's clan identity comes from their mother's family. So, to clarify, Jaime's Grandma Eleanor is not the mother of her mother but the sister of her mother's father. And what Jaime is describing when she talks about sheep camp is the Navajo practice of passing down traditional ways—kids spending time with their elders in the Navajo Nation, learning how to herd the sheep at the heart of their culture.

"It teaches you a lot about how to manage things when you're young," Jaime says. Mostly it teaches Navajo children where they come from. Traditional Diné culture reveres the sheep, holds that to care for sheep is to live in harmony with the earth. Jaime's family keeps these values close. And like many Navajos who traditionally raise sheep, they eat the sheep in celebration, use sheep fat in ceremony, and some spin wool for yarn dyed with herbs, then weave it in the Navajo style revered for its unique geometry, an economic mainstay.

Jaime grew up a short drive away from her Grandma Eleanor in Cameron, Arizona, in the Painted Desert. It's a place of less iconic canyons in the southwestern corner of the Navajo Nation, where blue sky meets red rock in stark lines, striations riding siltstone and mudstone and climbing out of the dusty earth in every hue of red and lavender. This corner of the world is rich in ways other than the commercial ones, with a deep cultural heritage and long traditions, though with patches of cruel economics. The great swath of land sits at the four corners of Utah and Colorado, Arizona and New Mexico in a chunk the size of Ireland. A century and a half ago, the Navajo entered into a treaty with the United States, which, among other provisions, ceded much of these original homelands to the Diné in a permanent reservation.

Since Jaime was a young girl herding sheep by day, hair over shoulders in one photo, dark eyes shining, she has had few departures from this life and this land—stints in Flagstaff about an hour away. Even now, when she is a student at Fort Lewis College in Durango, Colorado, this land never really leaves her. "Every time I go back there, I feel like I've just been recentered. And I just become connected to the Native American part of myself, even though I'm kind of usually just at home, watching TV." Being on the reservation, being able to talk to relatives, to be in that country with the people who have only ever been there with her, it's a way of coming home that Jaime feels acutely. The thought of losing it to the brutality of climate change—it is deeply, profoundly unsettling.

She describes this possibility not like a relocation, like a regrettable move for a job or for school, but like a dislocation of the soul, like the cutting of a person in half. It is one thing to move. Jaime has done it before for college. But home is home. And if she is forced to never live there again, or even visit, because of the way this land is remade by climate change, uninhabitable, "I think that's when all hope is lost. I think then once we can't live where we've been living for thousands of years, our culture is completely gone."

That the Navajo Nation has become a kind of ground zero for climate change is in the realm of the scientifically indisputable. The USGS has spent years measuring its unique combination of less rain, less snow, and climbing temperatures. Scientists have found that few springs still flow on the reservation, and that the Little Colorado is 10 percent of its former width in parts, down to 100 feet wide in Cameron from a former 1,250. At Many Farms Lake, a fishing lake two hours away, the water level has fallen from 16,000 acre feet to a few feet of silt in a muddy bottom. Snow depth across the Navajo Nation has similarly decreased by two-thirds since the 1930s, from an average foot of snowfall to 4 inches. Vegetables like traditional corn don't grow well or at all anymore. And the groundwater loss one can see in the Little Colorado is everywhere. The Navajo Nation's four main wells are an ongoing bustle of need, so much so that some people arrive as early as 3 a.m. to get the water that tends to dry out by 8 a.m. Or they travel longer on hard roads in search of water.

Scientists looking for facts about how these conditions have changed, and when the Navajo Nation got so dry, have found them recorded both in documents and in the memories of Diné elders. Both point to the same year, 1944, as the last in which the climate was wet and the snow accumulated in drifts, burying horses to their chests, with rain falling daily in the wet season. Now snow melts quickly, the storms too brief and infrequent to replenish the dry rivers and streams. There is no record of any year without snowfall before 1982, but between 2002 and 2011 there were eight years of little or no snow at all.

Rainfall has similarly declined since the 1990s so that the climate has been one of profound drought ever since. In 2009, for example, it rained only 3 inches. And the summer monsoons that used to make for a steady rainy season now come in intense, massive storms that further damage the land. All this means that where people once grew gardens and fruit, and farmed crops like corn, there is little but dry sandstone to usher in the planting season.

In Jaime's lifetime, the nineteen years in which the Little Colorado has shrunk, the world of wildlife around the river has shrunk with it, reduced to lizards. She describes how traditional herbs are getting farther away. Navajo tea, for example, a plant with a tiny yellow flower that grows in the desert, is one she remembers finding easily on walks near the Grand Canyon. Now, she says, you have to find a vendor or someone who has traveled to harvest it. This is a compromise of sorts. And it is the same with all the other herbs. Like the bitterroot for guarding against bad spirits. Or herbs used just for ceremonies, for traditional foods, or in prayer. In times past, Jaime's elders didn't need to have stashes of such things. The desert offered the plants and people used them. Today, these herbs are things to be hunted in markets or in long walks across the land like the elusive prey they are.

Navajo leaders troubled by these trends have undertaken a robust effort to do all they can to halt them. The nation's climate adaptation plan includes everything from deploying renewable energy to reviewing and revising who has what water rights and exploring ways to recycle water. Where headwaters and wetland habitat can be restored or preserved to recharge aquifers, the tribe aims to do so. And critically, the Navajo Nation also plans to halt the use of groundwater for mining, something that's been done in the past to tap the nation's rich coal and uranium deposits. Even things as rote as communicating about climate change and ways to fight it are part of this work, with education plans for 4-H groups and presentations in every district, where regular meetings about livestock and grazing tend to attract a crowd.

Now, wildlife technician Janelle Josea's job is to knock on doors, visit people. This includes the people who live deep in the desert and tend their sheep, speak Navajo, and don't much think about sending their kids to school. Her primary job is to educate, armed with booklets and PowerPoints, a mission conceived in 2014 when the Navajo Nation started work on its climate adaptation plan, holding meetings in each of its districts to gauge what people were most concerned about. And as a first course of action, it falls to Josea and three colleagues at the Navajo Nation Department of Fish and Wildlife to tell everyone on the reservation—about 175,000 people—what the plan is.

"There are a lot of people who are lacking this knowledge. Or even just knowing what climate change is," she says. "We really want to get this out and make sure our community understands it, and that climate change is real and it's happening now" and that the tribe is doing something about it.

Drastic changes to reverse these trends could occur worldwide if the United States conducted the very same analysis and planning that the Navajo are already undertaking. New water allocations. New grazing plans. Waste management programs, and revised timber and land management practices, too, along with the powering down of coal. It's not rocket science. It's arguably much easier than rocket science.

These days, the Navajo Nation's climate adaptation strategy calls for livestock management—outlawing the use of studs and artificial insemination, adopting horses off the reservation, enforcing these laws, and passing new laws to protect the enforcers. Some of these ideas run roughshod over old ways. Over just leaving well enough alone. And these tough conversations about having to manage resources, manage animals, they're the ones Josea and her co-workers need to have.

"Whatever works for us, great," says Josea. "If it doesn't, we can change that. That's why getting experiences and any kind of feedback from our community members as far as living conditions—that is helpful."

The Navajo Nation, like the United States, must still wrestle with its own fossil fuel dependence. Though the reservation has a long relationship with coal, and the tribe receives millions in payments from coal, gas, oil, and timber sales, the nation also anticipates the closure of two of its four coal-fired power plants by 2025 and recently closed a coal mine. While that's good for climate change, it brings the same economic hardships that befall other communities that depend on fossil fuels. Hundreds of lost jobs are no small issue in a place where 40 percent of the people are already living below the poverty level. The tribe anticipates $35 million in budget cuts in 2021, and it isn't clear whether renewable energy can replace what income has been lost, though the tribe has issued a proclamation to make developing it a priority.

In this way, what the Navajo Nation can do, it does. One-third of the reservation's population is younger than eighteen, and tribal leadership takes seriously its obligation to look out for its youth. But it can't restore nature's balance unless the rest of the world helps. These issues, they're part of bigger trends that can't be battled on the reservation alone.

Case in point: intensifying wind makes it hard for native plants to take hold, which makes it almost impossible to keep sand from accumulating, drifting, edging ever closer to homes and roads. Already there's a new kind of dust rising up, covering things in dunes. Scientists say it rises out of the dry riverbeds, like the dry expanse of the Little Colorado, and roves when no plants can take root to hold it. In fact, nearly a third of the Navajo Nation is covered in roving sand dunes now. The USGS has been trying to calculate how these dunes move as they do things like swallow up roads, fill horse corrals, and advance on people's homes, a new Sahara looking for purchase. The sand destroys farmland and suffocates plants. It covers rangeland, too, so there is less room for animals to graze.

These facts and the fact of water loss have turned conversation to reducing livestock on this reservation where sheep once ruled the economy, with woven textiles and millions of pounds of wool being brought to trading posts. Now, people in some parts of the Navajo Nation have to guard against dust pneumonia when the wind blows, hunkering down in conditions like a modern-day Dust Bowl, and there is talk of whole communities being relocated out of the way of the sand.

Such sand is not just a threat to the Navajo but also an accelerant of problems in other parts of the West. When the sand drifts, it ranges as far as the Rocky Mountains and lays a dark silt on the snow there, causing the snowpack to absorb more light and melt faster than it used to.

"I think a common misconception about climate change is it's just global warming so everything is just going to get warmer," Jaime says. "But this is a whole planet. And if something changes, it's going to affect something else."

Growing up here, Jaime learned about climate change when she was nine. When she joined *Juliana v. United States* at fourteen, she did it for the animals because she worried about what climate change would do to them. But Jaime worries more today about what climate change will do to her and to other Navajo people like her. Sometimes these concerns—about harms to the animals and harms to the Diné—are more or less the same concerns.

The summer of 2013 underscores what she means. That year the drought lasted into what used to be the rainy season. And the wild horses that roamed free, many were dying of thirst. There were reports of horses with ribs sticking out, horses found dead at dried-up watering holes, and horses who tried to kick their way into watering tanks for a drink.

The horses were a new kind of problem, generally. The land was already overtaxed by drought, and the horse population was meanwhile growing larger all the time. The Navajo had lived peacefully with feral horses for decades, but with the drought, they had to confront the reality of the land not being able to support the horses along with all the rest. The Diné were having cultural clashes over what should be done, whether the horses ought to be rounded up and adopted off the reservation or sold. Such talk collided with traditional values, like the value to just let the horses be. But no matter their feelings, people cared about the horses and did not want to see them suffer. That summer, the drought in the central part of the reservation was especially unyielding. In one report, the *Navajo Times* relayed the tale of a herd fighting for a wet patch of mud where a faucet had dripped, their bodies queued up so thick it was impossible to tell until after they were loaded into trailers, too tired and malnourished to fight, what the commotion was even about. Only after they were cleared away did the tiny spot of wet ground show.

A few years later, the spring of 2018 brought drought so severe that Jaime remembers it more starkly than most. The winter was drier than ever before, the spring runoff sparse, and there was no snowpack to feed the watering holes where the horses and livestock went to drink when the rivers ran dry. "In one part near Cameron . . . there's like this little flat part that absorbs a ton of water," Jaime says. She's talking about a stock pond where animals have long come for water in the worst of times, the water brackish and left to the wild. That year it was a sliver of itself. And what water was there was being swallowed, slowly, by the dusty soil underneath. The sparse water on dry soil—the two mixed like porridge, not like something you could drink. "It just became quicksand. Right in the middle of it, there was just a small pond, a little puddle basically. And because there was no other water in the area, all the wild horses around there came and tried to get the water."

More than 110 horses died in the mud, maybe more than could be safely counted after others slipped deep underground. Their bodies were found in a ring around the puddle, their coats caked in hardened earth from struggling for anything to drink. The ground around this muddy circle was dusty, held the bones of dead horses past, and later it was encircled in barbed wire so officials could move on to what came next: excavating bodies to deter scavenging birds and dogs, avoid disease, and clear out the smell.

"It's just kind of rough seeing things go down that we have no control over. It affects everyone," Jaime says.

For the rest of the summer, people hauled water for themselves and for the horses too. Rescue organizations sent hay and plastic tubs to make safe troughs. In another year of ceaseless drought, filling the tubs would become a new kind of normal.

These conditions horrify Jaime. And she worries most for older members of her community, like her grandmother.

"It's just really strenuous on the elders, who are the main people who still have our culture," Jaime says. She explains how many still live alone in the traditional earthen homes, tending their flocks of sheep and horses in the vast empty parts of the desert. They haul their own water, gather their own plants; some farm their own food. They face a tougher job as water gets scarcer.

Advancing age already makes life difficult for Jaime's Grandma Eleanor, older than eighty and living alone with sheep and a couple of horses. Jaime says her grandmother sometimes spends hours driving for water. She used to just drive into Cameron, but since there isn't enough water in Cameron for everyone, the wells run dry and she has to go to Tuba City, another half an hour away, or another hour to Flagstaff, sometimes even farther to Leupp.

Because the elders are the keepers of the old ways, the water gone out of the land can spell the tradition gone out of it too. If elders are no longer able to live a traditional life, there is less of a way for the younger generations to learn.

For Jaime, for many Diné, this learning is her spiritual core. Sheep camp was a summertime ritual, and though there was room indoors, she and her cousins often slept in the bed of their Grandma Eleanor's pickup truck, armed with BB guns. "She would mainly do it just to make sure that we could keep a lookout on the sheep, because summer is when the wolves just go crazy—or the coyotes," Jaime says. "It was really fun . . . if it was just a coyote, we would just yell at it and throw rocks, we wouldn't have to use the BB gun." Actual encounters of this sort were rare. Less rare: the nights in the pickup, tucked in blankets with cousins, the sky a cocoon of stars.

Preserving land and culture, preserving spiritual practice, these are the kinds of things Jaime gets to point out when she feels inclined to talk to press about why she is a plaintiff. That doesn't happen often. Life on a reservation didn't really prepare her for how public, how social being a *Juliana* plaintiff would turn out to be. "There's no major news outlets. There's nothing crazy going on." Life in and around the Navajo Nation is a thing that moves to its own clock, and it seems to be a slower clock than the one running the news cycle.

In the stints between canceled hearings, when the only job to do is to give talks or talk to the press, Jaime finds herself shrinking back. She is not outgoing. Not particularly talkative. And she finds it hard to figure out what to say. Crippling, even, when the interviews stack up, one after another in short spans of time. She has the challenge of being an introvert. Of recharging alone and struggling in situations when there are just enough people around to make things feel like a circus. It isn't in her nature to run toward such things. And while she has friends among the plaintiffs and likes them, she finds herself drifting away after gatherings and meetings, avoiding the press and going back to her own quiet corner of the world, where things move at her own pace in a land she understands.

"At the same time, I think because of my culture, this is the reason why I am doing it. The whole reason," Jaime says.

This specter of displacement, of cultural and spiritual loss, is already a reality for millions of people worldwide. It is more unusual for Americans than for people in other parts of the world. Many who face displacement elsewhere are also from Indigenous communities, with nations in Africa and in Asia, as well as island communities home to some of the most immediate climate impacts. In this way, Vic Barrett will tell congressional leaders in September 2019, developed nations are foisting the worst of the impacts of climate change onto people of color, often people who are the least able to bear it.

Even though 25 percent of the carbon dioxide that's accumulated in the atmosphere since the industrial age is attributable to the United States—and the government admits this, from its own filings in the *Juliana* case—it is poorer communities, often communities of color, that face the worst impacts.

To explain this to a joint meeting of committees on climate and foreign affairs in the House of Representatives, and in an op-ed in the *Guardian,* Vic will point to his Garifuna people, the Afro-Indigenous community from the Caribbean island of St. Vincent first displaced by British colonizers in the eighteenth century to eastern Honduras and Belize, and now in danger of being uprooted again by sea level rise from climate change.

The Garifuna are a community known for contributions to percussion-driven music and fish-based cuisine, who bring island flare to a Latin world, who, like many migrant communities, live with a kind of day-to-day otherness in ordinary Honduran society. Which makes them especially vulnerable as another group of people facing more intense and frequent storms in a land set to be underwater in a few decades. Already, the offshore coral reefs central to the fishing culture of the Garifuna are being eroded by the bleaching effects of rising temperatures and nutrient imbalance. Vic grew up in New York. But his family has lived on oceanfront land in Honduras—Vic's inheritance—for generations. That, too, is about to be underwater.

"Frontline communities around the country and around the world are already feeling the effects of the climate crisis, from the dispossession of land to the grave public health threats that are disproportionately affecting myself and other young people," Vic tells congressional leaders. He's at a microphone and flanked by young climate leaders, Greta Thunberg on his right, a row of *Juliana* plaintiffs at his back. "These frontline communities are made up of people who look like me. Young. Black and brown. LGBTQ. Indigenous. Identities which place them at significantly higher risk to experience the impacts of climate change than the general populace, due to their marginalized status in our society."

These factors make climate change a profound social justice issue, Vic will write. An issue, he notes, that moves talk of climate justice past intergenerational inequities and to questions of poverty and of race. It's an issue that provokes a central, unsettling question about fairness and American culpability that is perhaps more distant from the national conversation about climate change than it should be: Who benefits from a fossil fuel economy, and who pays? Not just abroad, but in America too.

CHAPTER THIRTEEN

"I Will Be with You in the Streets"

It's Christmas break and everyone is tired of everything, including the town they live in, so Kiran's band goes on tour with a group of others. The band is called Geophagia, which is the medical term for people who eat dirt. They stop in Portland on a blustery night when the rain is coming down sideways and play late at the Twilight Cafe, about the only place of its size that still has a decent stage and hasn't been turned into apartments yet.

The cafe is in a part of town that will probably never turn up in a *Portlandia* episode. It's in the section of southeast Portland between the nouveau-riche high towers, ever encroaching from the west, and the area to the east that seems an homage to the car culture of old—cinder-block buildings smattered along a defunct highway, the too-many lanes a pedestrian terror, lined with Burger Kings and Taco Bells and other places with parking lots.

Inside, the Twilight is a place of hardiness and endurance. Of tall boys and appendicitis, face tattoos and banjos. It is red-and-black-painted with black upholstery, a black-painted bar and stools. Behind the bar is a collection of televisions. One of them is playing a weird movie, another an underwater documentary, sports on a third, and the fourth broadcasts what is happening on the stage to the bar patrons marooned on the wrong side of a wall in between. The place is shaped like an L, and it is packed. People with short bangs, denim jackets, knit caps and beards, fuzzy scarves. Almost everyone is wearing some shade of black plus patches, spikes, sometimes chains. A guy with a green knit bandana sticks out, adorably, like a deer in a hunting vest. Somehow everyone is both impressively young and impressively youthful, and the show encapsulates everything young people ought to be worried about.

Kiran is on stage when I arrive, act three of six, playing the mandolin and singing while their partner plays drums in black lipstick. A singer in a leather skirt alternates between an accordion and a melodica. There's a stand-up bass, a guitar, a violin. And everybody is wearing a dark-colored T-shirt. Behind them, the mural on the wall is a simple black-and-white spiral, one that looks not like the 2D paint that it is but like a tube that falls endlessly inward, like it might swallow the band and concertgoers along with it. The concrete floor in front of the stage is full of people. So many they are spilling into the handful of booths at the bend in the L; and the sound booth, which is tucked into a corner not far from the stage, is dangerously mobbed with dancers. Above my head, Kiran's image is being miniaturized in a cell phone while it records, the backdrop spiral a halo recast in purple stage light. They're singing to a tune that moves between an accordion waltz and a string riff, with a soaring violin that will soon end in screaming.

We are nonsensical mindless identical
The way things are going soon all will be lost
I don't need to be hopeful for the future
Standing here watching all this refuse floating by

Kiran has just returned from Switzerland, where they met with the ecumenical patriarch of the Eastern Orthodox Church and spoke at the World Council of Churches in Geneva. These are the kinds of things Kiran is invited to do because of the lawsuit. There's a photo of this meeting on Facebook, Kiran in a gray suit, at attention to the robed high priest, "the pope of the East," Kiran calls him. Right now, this recent meeting seems supremely ironic, since Kiran is looking quite secular—hedonistic, even—in more usual clothes: a studded black denim jacket with a Subhumans patch on the back, black jeans, and long belt with a chain.

The United Church of Christ (UCC), for which Kiran was an emissary abroad, has long embraced this dual ethos. Kiran's mother used to be a minister in the church in Eugene, where the UCC teaches that Earth is a God-given home, a thing to love and to steward. That the church embraces this idea helps explain how Kiran's deep commitment to the environment was fomented young. Kiran thinks not enough people know about the UCC and in another conversation told me: "Basically, everything everyone thinks is bad about American churches, they're doing the polar opposite." Nonpracticing though Kiran is, the UCC is still a touchstone: good people who do good work, if not in the grassroots way that Kiran most relates to, the way that is all around us tonight. This scene at the Twilight, it is another kind of religion.

The band wraps up and Kiran takes a seat next to me on a stool. It's briefly quiet between acts, while Kiran tells me that worry is dimming the glow of the tour.

When Kiran first got the text that their deposition was canceled nine weeks ago, the feeling was only relief. Life at the time was a binding itinerary of classes and exams at the University of Washington, the pending deposition, and shifts at work. Kiran desperately needed a weekend off. But now that school is out, the nothingness of it is making for nerves. The deposition still has not happened, and no one has heard from the Ninth Circuit about what, if anything, comes next. And while tour is a nice distraction and it's a relief that Kiran can go from town to town—Tacoma, Olympia, Grants Pass, Eureka, Oakland, Eugene—in this folk-punk bubble in which everyone thinks alike, feeling amazed that the bubble is everywhere, there is also anxiety that the case has stalled. Kiran is not sure what it means, or if it means something bad. There is a conference call with the communications team soon, and meanwhile Kiran is "trying to chill." This week is to be a week without email. Kiran forgot to turn on the autoresponder but their partner says it's okay to take a break. Forget about it.

A man with a face tattoo is on stage now, playing a banjo and alternating feet between a tambourine and a bass drum made from a suitcase. There is a guitar player on the man's left, and both are adding vocals so throat searing it seems laryngitis is perhaps the best outcome. The room is full of sore throats anyway. The bartender—who proudly claims he has only ever missed one day of work, and only after five days of ignoring what turned out to be appendicitis—is attentive to the winter colds. He has a simple syrup made from ginger and honey and is serving splendid hot toddies, actually troubling himself to warm the whiskey and the mugs.

Kiran is driving tonight and not drinking, partaking of the culture instead. A feature filmmaker recently suggested that Kiran could play themselves in a film, and Kiran is kind of okay with that. The thought of this film is settling the other nerves about the stalled court proceedings. It signals that despite the cancelation of trial, the case is holding the media gaze, at least for now, and gives Kiran hope that the *Juliana* mission is not over, even if the case is done. There is still momentum, and the *Juliana* twenty-one are still building attention for the climate movement. If the vehicle for that message turns out to be a feature film instead of a courtroom spectacle, it is not the hoped-for outcome but it is also not the end. Just another platform, another way to spread the word, Kiran says.

The Window-Smashing Job Creators, a band that will be more widely known within the year, takes the stage. Their lyrics include motifs about things like wanting to read Karl Marx in an Applebee's and then burn it down. And about how you cannot change the world unless you are willing to talk to people you don't know. Or about how there is no left, and other politically fretful ideas that at one point cause someone in the audience to shout out "Bernie!" and a number of others to thrash around—half mosh pit, half square dance. The saxophonist is playing through a ski mask and the bass player is wearing a red bandanna like an Antifa. One moment there is a quick guitar riff that sounds like a nod to the fifties, then a cover of a Beastie Boys song that used to be about only wanting girls but is now about only wanting bread. There are unflattering lyrics about the Portland police interspersed with clapping, and a guy in a Santa hat moves along the rim of the dance floor, happily clapping along.

For the last two songs, the Job Creators invite a fiddle player named Lightnin' Luke, who climbs the stage in a navy button-down, a black sport coat, black jeans, and a black fedora. He is bearded and very on his game. With an accompanist of sorts, whose name I do not catch, the duo becomes the last act. The guy who isn't Lightnin' Luke plays a bass kick drum and a tambourine, plus a steel guitar and a kazoo with a weird little speaker on it that makes it all look very steampunk. Kiran is off dancing then, finishing the night.

This is a tiny club. Tomorrow it will be home to another few bands playing another tiny show. But these kinds of grassroots movements in music, the ones that pull from the world with a force that starts to resonate, they are the ones that rear up. Like Nirvana. Like the rise of nineties punk. And if not, what then?

Julia Olson's words from October come back to me, from the day the trial didn't happen, when she spoke passionately from the courthouse steps, promising that if the courts would not hear the youth on climate change, "I will be with you in the streets." If Kiran can't settle for a feature film while the trial gets farther away, can't settle for an opportunity to just spread the message however they can, what next? If we see them in the streets, burning the proverbial Applebee's, will we be surprised?

Within days, *Juliana* stalls out for real. The details are legal ones, again head-spinningly technical, but the gist is that on December 26 the Ninth Circuit grants the interlocutory appeal offered by Judge Aiken the month before. Essentially, what this means is that the Ninth Circuit will decide if the *Juliana* case should get a trial at all, after reviewing the government's arguments that the case should be dismissed. The effect is that *Juliana v. United States* flatlines until a hearing that's ultimately set for June 2019, six months away. And while the Ninth Circuit could move quickly from there, the change also gives the US Supreme Court a pretrial opportunity to weigh in. If that happens, it will mean the trial could begin as late as early 2022. And with posttrial appeals in the mix, it also means that the *Juliana* plaintiffs could be six years from the finish line, or halfway to the IPCC deadline for taking action to halt climate change.

For a while, things are just quiet. And the possibility of this spectacle drawing national attention to climate change seems to be over. Without the trial to feed the zeitgeist, the imminent end to a habitable planet seems to have a tough time pulling rank in the long list of things to worry about in America. Increasing violence, rising radical nationalism, not to mention the day-to-day tangle the young are having with the cost of student loans, of housing, of working three jobs and wondering where it is all headed, these are the things commanding the conversation. Though Democrats set a House agenda to include climate change, and talk of a Green New Deal is picking up steam, it's hard to know whether the glacial pace of governance will be faster than the actual melting of the glaciers, about which the news is daily getting worse.

This new round of waiting, it's something Aji Piper has been doing for years. A plaintiff in the *Juliana* case, Aji was a state-level plaintiff in the Washington litigation too. And in the state case, *Foster v. Washington Department of Ecology*, the kids won, sort of. A judge didn't force a new clean air plan to cap and regulate greenhouse gases. But she told the state to complete a clean air rule that had been stalled. Predictably, the rule did little to reduce emissions later. Plus it was largely overturned by another court. So the plaintiffs sued again in February 2018 in a case that was dismissed. Now an appeal is pending. Which is why Our Children's Trust has been so keen to involve the press and public in its work, and build coalitions with activist groups—because its leaders recognize how key public pressure is to seeing the government actually do what the courts order. *Brown v. Board of Education*, the landmark civil rights case that desegregated schools, wasn't much different. It was successful because people were in the streets. There was unrest. Society demanded the government heed the court.

Aji Piper has long wanted similar action on climate. Something bigger than the court. He told me as much on a walk in Lincoln Park in Seattle not long before, while I struggled to keep up with both his lengthy strides and the way he strings creative thoughts together, sometimes in non sequiturs. Aji is unique in many ways, maybe chiefly in his ability to preserve a kind of life aesthetic that eludes even diligent people. He is bored by social media, disdains the unyielding dopamine quest of likes and faves. He likes authenticity. Workmanship. Originality. What is new, inspirational, genuine draws him in. In this conversation, Aji easily reminds me of a young Abbie Hoffman, and not for the big hair. But because Aji spends a lot of time thinking about where performance and activism converge, and how they best make for tangible results.

Here's the story that inspires Aji the most, he says, walking through the maple leaves, fog lifting off Puget Sound below: the Rainforest

Action Network (RAN) Toronto 2009 Nixon campaign. And no, not that Nixon.

It was an advocacy campaign that, through a series of public spectacles, steeped a city in intrigue. It began with a simple poster. No branding, a plain design. Just: "Please help us Mrs. Nixon." Activists had earlier forged a partnership with a local garbage union in Toronto, promising to strike with union workers if they would leave the poster alone. The union agreed. So RAN blanketed the town. Organizers covered walls and transformer boxes, light posts, and everywhere else until the city could not get away from this poster.

It was a campaign of mystery. Of theater. Because few could identify who Mrs. Nixon was, it entranced Toronto. She was not a politician. Not a well-placed executive. She was not a public figure at whom people could readily point. So the draw was not the plea, but the question: Who is she?

Activists dropped banners over highways when the traffic was thick. As people sat in their cars and waited, "Please help us Mrs. Nixon" would unfurl overhead along with the dial frequency of a pirated radio station. When drivers tuned in, there were stories from people who lived downriver from the tar sands, people who talked about mining's effect on their water and their lives. RAN also had people riding the elevators in the office building of the Royal Bank of Canada, stealthily distributing flyers and creating buzz. "Please help us Mrs. Nixon."

The rough contours of the ask were visible enough—tar sands, bad—but the target, Mrs. Nixon, kept curiosity stoked for weeks. After a month of provocation, the reveal was this: two Indigenous women climbed the flagpoles outside the Royal Bank of Canada, then dropped a banner between the Canadian and the provincial flags. It read "Please help us Mrs Nixon.com." And on the website, a video described $2.3 billion in tar sands investments from the Royal Bank of Canada over two years. The video also described RAN's ask: that the public urge the Royal Bank to phase out tar sand investments.

More than three thousand people clicked through the site to write twelve thousand emails to bank executives, enough that insiders later said the effort crashed the bank's servers. Ultimately, the campaign forced a two-year inquiry into the legal risks associated with tar sand investments for Royal Bank shareholders. Mrs. Nixon turned out to be the wife of Gordon Nixon, the bank's CEO. She was a committed environmentalist.

"It's intensely creative. That's something you don't see all the time," Aji says, smiling, wistful, deep dimples in his cheeks.

Through the park's trees and greens, he wonders aloud what similar thing can be done to urge climate drawdown. Can he gather enough people to stomp, register an earthquake on the Richter scale? It happened once at a Seattle Seahawks game. Whatever it is, he knows it isn't a march. Isn't something that has been done before. "For, like, fifty years people have been chaining themselves to things, standing in front of things, blocking things, marching and yelling with signs. It's time for something a little more shocking," he says.

It's not that he thinks *Juliana v. United States* won't be the thing that stops the clock on climate change. Won't still be the galvanizing force this nation needs. But any time between now and then? It's time wasted, time in which something else is bound to happen.

After this news from the Ninth Circuit, Kiran's music follows a trajectory of punk. Which is to say that what started out with acoustic instruments, rebellious in its message, becomes rebellious in an aesthetic sense too.

Maybe this would have happened anyway. After all, Kiran says the first band that caught their attention was called Crass. It was the first band, at least by Kiran's standards, to play what can be described as anarchic punk. The band lived in a commune. Talked seriously about politics. Went to protests. And left a legacy that was as much word and deed as it was sound. So while Kiran has listened to that punk icon the Sex Pistols, and knows that the origins of folk punk date back to guys like Woodie Guthrie and Bob Dylan and Utah Phillips, Kiran's own punk life started with the punk that got harder and louder until that point in the nineties when punk almost sounded like metal, when it started to lose some of the force of the lyric that first drew Kiran in, then turned back to folk punk.

That's what Geophagia is. Folk punk. A few steps away from straight folk, which today is a long way from Woodie Guthrie, played by people who are not all that political anymore, maybe because of the commercial success of their sound. But not terribly long after the show at the Twilight, Geophagia will take a break. Maybe a permanent break. And Kiran will join a new band called Pickax with some of the same people and start playing music that sounds like all of them seem to feel. The mandolin is no more. It is hardcore punk all the way—and power-violence. "Just drums, guitar, bass. Distortion turned all the way up. I don't get to sing anymore, I just scream. And we play a hell of a lot faster than we ever did before," Kiran says.

They start trying to figure out how to write protest songs. Start doing benefit shows, one to stop pipelines from crossing Indigenous lands in Canada, and another for Rojava in northeastern Syria, the stateless feminist society that fit millions into an ecologically sustainable system.

Kiran acknowledges that this shift is "going alongside the shift in the way that I'm feeling about the movement." This pent-up energy that comes from the environmental angst, it flows with an increasing call for response, for action, urging people to join the front lines.

CHAPTER FOURTEEN

"The Constitutional Question of This Century"

A funny thing happens: the media eye keeps staring. Maybe the machine isn't satisfied, after all. Maybe the tease of the trial was just too much, the beast still needing to be fed. Maybe it's just that some news shows, like *60 Minutes*, were already midproduction when trial was canceled and decided to go ahead anyway. Within the next few weeks, the plaintiffs will be prime time, Levi walking on a beach with correspondent Steve Kroft, and the whole gang scattered on the courthouse steps again, this time for a *Vogue* photo shoot, careening toward a feature in *People*. The momentum of the almost-trial is still rolling.

That sense that Aji has that there needs to be something bigger, or that Kiran has that perhaps winning the pop culture war will be the victory after all—both seem to be right. There is a noise in the world that is coming from the youth, a kind of siren. It is getting louder.

Halfway around the world, fifteen-year-old Greta Thunberg has been sitting outside the Swedish parliament protesting inaction on climate change on Fridays. She's been on a parallel track with the *Juliana* plaintiffs this year, demonstrating since August when the *Juliana* twenty-one were doing depositions and trial prep. By December 2018, she's become popular enough to be invited to speak at the UN's climate conference in Katowice, Poland, where she delivers the lines that will make her what *Time* later calls an "avatar" of young activists everywhere: "You say you love your children above all else, and yet you are stealing their future in front of their very eyes. We have come here to let you know that change is coming, whether you like it or not."

By then, tens of thousands of young people across Europe are skipping school to join her Friday protests, and in the beginning of 2019, thirty-five thousand children attend a single protest in Belgium. Then a brigade of teens leads twelve thousand on a march into the Netherlands to The Hague. Two young women from Brussels next make a pleading video that draws more than one hundred thousand youth to a march on the capitol there. Then a twenty-two-year-old Berliner answers the call and leads a climate strike in Germany.

In the United States, a trio of young women begin organizing the US answer to this movement abroad, US Youth Climate Strike. Young organizers from Australia, Uganda, Britain, and China also join, orchestrating their own strikes and planning larger actions too. Solitary pickets of capitol buildings and outposts begin to grow into crowds all over the world. As youth assume the leadership the adults have abdicated, they stall traffic in the streets, halt buses, make signs and slogans, and recite chants asking why they ought to be in school anyway. They don't care about being the next cogs in the wheels of a civilization on a disastrous course. They want change. And they reason if they stop showing up at school, the effect on the adults will be that they have to start paying attention. These youth tell their leaders, their parents, and their teachers to listen up as they stump for global climate remediation. And it increasingly seems that some will get it. In nations across the globe, action influences elections that influence policy.

They call it Fridays for Future, with some students boycotting school every Friday to protest global inaction on climate change. Talk of a nationwide climate strike hits the street, lights up social media, and the American answer to youth movements abroad is set for March 15, 2019, a Friday, where a school boycott and climate march is planned on the Capitol lawn in Washington, DC, with solidarity events planned in schools and towns across the nation.

Climate organizations founded for and by youth meanwhile explode into policy debates, voicing support for climate legislation, shaming business leaders and policy makers with ties to fossil fuels, and skewering older adults for inaction. Groups like the Sunrise Movement and Extinction Rebellion, intent on raising the voice of youth and disrupting business as usual, follow in the path of organizations like iMatter, Earth Guardians, and Zero Hour.

Fitting, then, that Jamie Margolin, the seventeen-year-old co-founder of Zero Hour who led a climate march in Washington, DC, in July 2018, one event among many to inspire Greta Thunberg, has already tried her luck to force climate action in the courts. She was Aji's co-plaintiff in his second case against the State of Washington, the case in which the plaintiffs won a clear air rule but were never able to turn it into the climate recovery plan they sought. Aji was there the day Jamie first conceived the 2018 climate march. Back when it was a chore. Now, Jamie says marching doesn't feel so frustrating anymore. The fever to do something, anything, runs deep among many, and the urge to do whatever has to be done, and do it by the thousands, is no longer a monumental task for just Jamie, her colleagues and co-plaintiffs, and a slimmer number of other youth.

As young Americans plan to hit the streets, Jamie turns her attention elsewhere. On February 8, 2019, the *Juliana* plaintiffs file a motion with the Ninth Circuit asking the court to freeze fossil fuel development on public lands before their case can be heard, a prospect that would interrupt an estimated hundred such projects around the nation. Our Children's Trust asks Jamie to leverage Zero Hour to help, so Jamie recruits a batch of new youth volunteers, then bushwhacks among young Americans for support for *Juliana*. Pro bono attorneys craft a friend-of-the-court brief for Zero Hour. It urges the Ninth Circuit to bring *Juliana* to trial—and quickly. Then Jamie asks other kids to sign. NowThis News pitches in. So does a writer from Jimmy Kimmel and, again, Leonardo DiCaprio. The objective, Jamie says, is "showing this isn't just twenty-one young people suing the government," that the frustration the *Juliana* plaintiffs feel is being felt by young Americans everywhere.

"Our childhoods are being spent begging them to stop ruining our adulthoods, and our adulthoods are going to be spent dealing with the consequences of their actions," she says, this on the phone during a break from high school in Seattle, where she's a junior. "I don't want to be picking these fights. The youth don't want to be picking these fights. And honestly, it's exhausting to be in the streets all the time. . . . But people don't seem to want to listen to anything civil."

In about six weeks, Jamie and her cohort gather more than thirty thousand signatures.

The day of the US Youth Climate Strike march, the Capitol lawn smells like bark dust and twittering birds. It's covered in beach balls and parachutes and a hefty crowd of young people. The crowd chants for a while ("Hey-hey, ho-ho, fossil fuels have got to go!") and almost everyone has a sign. The messages range from the assertive ("Like the sea, we rise") to the ironically humorous ("I'm so angry I made a sign").

There are talks by organizers, passionate and breathless, and music by performers, heartfelt songs by participants too. There's a way to text your representatives to ask them to back a Green New Deal. And a solid showing by oldsters, including one man handing out leaflets and urging attendees to ban the bomb. It's a crowd of backpacks and tote bags, ripped jeans and knit caps, bike helmets and skateboards. Surprisingly, there are very few phones. Everyone seems to pay attention to the stage whenever anyone is on it. Those on it include, among others, the three organizers (Haven Coleman, Isra Hirsi, and Alexandria Villasenor), soul pop singer Rebel Rae, Congresswoman Rashida Tlaib (whose daughter is Isra Hirsi), and an eight-year-old self-described "tiny diplomat" who has to stand on a ladder to be seen.

Their ask is articulated throughout the day: a Green New Deal, an end to new fossil fuel infrastructure, compulsory education on climate change for youth, a national emergency declaration on climate change, and for the nation to reduce greenhouse gas emissions by 100 percent by 2050. In the world ahead, the youth also want government decisions to be based on science.

In the crowd, a young man sits on a parachute with a sign listing the ten hottest years on record. He is seventeen and tells me he has been alive for nine of them.

Later, organizers count school boycotts and rallies in forty-seven states. In another six months, they will do it again, this time joining at least six million others worldwide for the largest climate march in world history.

When it's over, Alex meets me for ramen and sushi burritos in New York. He was a speaker at the US Youth Climate Strike rally at Columbia, where afterward he interviewed Jay Inslee, then a presidential hopeful for 2020, on MTV News with a couple of other young activists. It was an interesting conversation, Alex asking what Inslee planned to do for rural communities left behind by globalization and automation in a Green New Deal. These types of invites—to speak to crowds at Columbia and interview presidential hopefuls on MTV—are the kind of invites he and other *Juliana* plaintiffs are increasingly getting. Alex has a wry sense of humor about this newfound celebrity. Earlier in the day, he says, a fact-checker from the *New Yorker* called him to confirm the details of an upcoming story. "The fact-checker asked me, with no trace of irony, if I was withholding my official endorsement of Inslee," Alex says. Then he just laughs.

He is about to graduate from Columbia and is working on his thesis during spring break. Working on a lot of other things, too, his brain on its unique overdrive while he considers what life after college will look like. He turned up to the restaurant just off campus wearing a collared shirt and a black peacoat, having just had three job interviews, and is juggling possibilities that include the prestigious fellowship he has applied to, beginning the ascent to law school, and a handful of potential jobs. Despite the pressure of these decisions, Alex is as upbeat as always, looking forward to a night out with his girlfriend to celebrate the interviews that seem to be going well. By this time, managing pressure is just a skill that Alex has mastered, like a lot of others. Since October, he has spoken to the United Nations about the *Juliana* litigation, appeared in media with the rest of the plaintiffs, and given his own interviews to the *New Yorker* and the *Yale Politic*. Some of the plaintiffs are set to testify before Congress later in the month, and Alex is thinking of joining them.

The stakes in the *Juliana* case are only rising. The Ninth Circuit is scheduled to hear oral arguments in June to determine whether it should go to trial. And by now, it is clear that the climate movement has changed—not just for Alex, not just for his fellow *Juliana* plaintiffs, but also for the nation and the world. Such momentum could prompt the courts to move faster. And we acknowledge as we sit, fussing over rice and noodles with chopsticks, how much has changed. I ask Alex what he thinks it is that is waking the world. Besides the budding iconography of Greta, whether the lawsuit has anything to do with it, or the new recruits in Congress who began the talk of a Green New Deal. I spitball a bunch of ideas. But Alex doesn't buy any of them. Instead, he says, unequivocally, "It's Donald Trump." I'm not sure I buy that, either, so Alex explains. "It's having a good villain. . . . I think every movement, every story needs a really good bad guy to motivate. This man is like the Darth Vader of global warming." Then he likens Trump to the James Bond villain Goldfinger. Calls him Orangefinger.

It is perfect.

In the 1964 spy film, a James Bond classic, Auric Goldfinger is a wealthy gold trader intent on destroying the US economy so that he can make himself comparably richer. He has a history of smuggling and has stashes of gold bars all over the world—Caracas, Hong Kong, Zurich, Amsterdam—which he intends to sell at triple the gains to a guy in Pakistan. The plan only works if Goldfinger can devalue US gold first. And he's in league with a horde of criminals who support this dubious plan. These villains meet in boardrooms, where they use fancy toys and technobabble to make the business of collapsing the government all the more official. Bond catches Goldfinger in the act when Goldfinger poses as a US military man, tries to kill the entire military, then runs away after being spotted trying to detonate a bomb in a gold vault. Later, Goldfinger dies by being sucked out the window of an airplane because he is fool enough to fire a gun on it. In between, he is obsessed with winning, enough to lie and cheat, and beautiful women are inexplicably drawn to him.

In this moment, I can think of no better analogy for the leader of the free world.

Congress seems to wake up, and suddenly even presidential hopefuls want *Juliana* endorsements. It isn't clear whether any of these politicians will change anything, but many congressional leaders understand they at least have to act like they are trying. Over the course of the next few months, the *Juliana* twenty-one become something of a stage prop. Aji testifies in April before the brand-new House Select

Committee on the Climate Crisis, then stars in a handshake video with Senator Ed Markey of Massachusetts. Bernie Sanders, again a presidential hopeful, makes videos supporting the *Juliana* plaintiffs, including talking-head segments starring Aji and attorney Andrea Rodgers, interspersed with shots of smokestacks and oil rigs churning, highways chugging like platelet-filled arteries, and clips of the American flag. By mid-May, Alex is a popular selfie on the Hill, posing with congressmen from Oregon and California, and Representative Raúl M. Grijalva from the soon-to-be-uninhabitable state of Arizona. He meets with Representative Dick Durbin, the Democratic whip from Illinois, and lands his own star-studded selfie with Alexandria Ocasio-Cortez. In June, Sheldon Whitehouse, D-Rhode Island, posts a photo of himself high-clapping Kelsey in a marble hallway next to an AED machine.

Then it is June 4.

The *Juliana* attorneys walk into a federal courtroom in downtown Portland with a rolling cart stuffed with case files. Our Children's Trust is throwing so much legal firepower at today's Ninth Circuit hearing that it's a task just to get all of the attorneys into chairs. A panel of three judges is set to hear arguments about whether the *Juliana* trial should go forward. With only a hundred seats in the courtroom—a wood-paneled place with the typical wooden benches—plaintiffs, their parents, and the press are seated first. Then the court staff begins the tall task of filling the remaining seats and a nearby overflow room.

The people who have lined up to attend this hearing, most of them are still school age. Many have no affiliation with the case, no affiliation with the plaintiffs. Their only affiliation is the issue. They are here because *Juliana v. United States* is about them.

In fact, there is such demand to see this hearing, and enough of an effort on the part of Our Children's Trust to make it visible, that the court has agreed to livestream the proceedings into nearby Director Park, an open plaza between a triad of towering buildings and a historic movie theater. In the park, it is bedlam. The stone ground is covered with supporters and activists rallying around signs to "Let the Youth Be Heard." They assemble under the watchful eye of twelve giant puppets—puppets of the last twelve presidents, the ones to have presided over the nation's lackluster response to climate change—as organizers rally the crowd in anticipation.

The plaintiffs, most of them, are of course here, wearing the combined expressions of excitement and composure that are typical in court. They are catching up, many not having seen one another in the months since trial was canceled. Miko and Jayden are in intermittent confab, Jayden having left Louisiana for New Mexico, leaving behind the oil politics for good. Jacob took a day off the farm and off school, having been admitted to a prestigious online program that MIT has begun to combat problems like poverty and hunger, geared toward nontraditional learners like young farmers. The Eugene high schoolers are seated together, having made it through their freshman year. And Kelsey, also back to school at the University of Oregon, is center stage, turning in her seat to face Vic and the plaintiffs behind her to check in. Everyone is in their finery, Isaac especially crisp in a black button-down, gray suit jacket. Nick, who has just finished high school and is about to begin studying theoretical math, confesses that he failed to pack appropriate socks, then comically flashes his ankles.

In these moments before the hearing begins—long before the plaintiffs emerge from the courthouse to the flash of cameras and a parade that will march them to the park, Levi riding on Nathan's shoulders and Kelsey screaming the news that today's hearing has just become a Snapchat filter—the courtroom is quiet, the tension thick. The Department of Justice has assembled its team of four attorneys, and Jeffrey Bossert Clark, clad in a gray suit, is poised to argue the government's case.

The room grows quiet. And in these last few moments, a courthouse staffer begins reciting "Jabberwocky" for a final sound check of the microphones: "So rested he by the Tumtum tree, and stood awhile in thought."

There is a soft, collective giggle from the youngest people in the room.

When it begins, Clark argues that the *Juliana* case is an attack on the separation of powers because it circumvents a lot of things. A bunch of statutes. Public comment periods. The Constitution, arguably, by not letting Congress solve the problem of climate change. If the case were allowed to go forward, he says, it would mean that any situation in which someone made a claim that their health and safety was in peril could be made into a constitutional case, skirting the usual administrative appeals and other required processes, the ones that Mary Christina Wood said were not working in the first place.

Judge Andrew Hurwitz frowns with his forehead, leans across the bench toward where Clark stands, and gives Clark a hypothetical: Say rogue raiders are coming across the Canadian border into the Northwest, kidnapping children of a certain age and murdering them. Say the White House refuses to do anything. Say Congress doesn't act. Can those people go to court to make it stop?

"No," Clark says. It just isn't what the judicial branch is there for, to deal with these kinds of exigent threats. That's why we have a president. Why the president can take the nation to war. The remedy in the murderous kidnapper scenario, Clark says, if the president and local governors aren't doing anything about it, would be to vote the president and the governors out of office. "It's not for the judiciary to take over," he says. However painful, to preserve the constitutional design of America, a few unlucky kids would have to die by the hand of rogue Canadian kidnappers.

Julia Olson challenges all of this. Next at the podium in a stiff dark blazer, wide collar and glasses, tiny by comparison, she casts the plaintiffs—casts all youth—as a class of citizens denied equal rights. They have a constitutional right to life, liberty, and property, she says. And because of the government's dependence on fossil fuels, the subsidies and permits it provides to industry, the policies that promote a fossil fuel economy in spite of science showing it is harmful, American youth have been denied these basics.

She clarifies she isn't saying that the government hasn't done enough. But that the government has caused this very harm. It has created a danger, the same way an errant cone on the highway can cause a car crash. The government has to administer the energy system, sure, she says. But it has been given the discretion to do that in everyone's best interest and, essentially, has screwed it up. "The scale of the problem is so big because of the systemic conduct of the government," she says. And when the government's conduct is shocking, when people are hurt, they are entitled to ask the courts for help.

When the judges ask whether the courts have ever done anything of this scale, have the authority to order the government what to do, Olson gives the legal answer first. She reminds them of the school districts that were ordered to desegregate—whole systems in whole states—and other cases like the prison releases that eased crowding in California.

Then she makes the moral argument. She says, "When our great-grandchildren look back . . . they will see that government-sanctioned climate destruction was the constitutional question of this century."

CHAPTER FIFTEEN

White Noise in a Land That Doesn't Care

Ten weeks later, Nathan Baring is standing on a boardwalk just above the waterline of a bog at Creamer's Field in Fairbanks, Alaska. All around us the trees are sinking, slowly being swallowed by water. There was always a lake here, Nathan says. But not like this. The water has crept into the boreal forest and swallowed an entire stand of birch trees, dead all around us, drowned. Denuded of leaves with their white branches reaching skyward, their trunks are submerged. The undergrowth of what used to be forest floor is dead now, too, hugging the trunks of the birches in a woody snarl of leafless bramble. Boggy grasses have sprung up, and the dead wood that's fallen to the wet floor is starting to grow moss.

This was already a marshy place, so parts of the trail are covered by boardwalk intended to cross shallow puddles or mud. Now, in places, the trees have fallen one into another, so that all around us they are leaning, teetering, ready to drop. It is not a place I would want to be standing in a swift wind.

This spectacle is called a thermokarst. Nathan asks Siri to explain to me what that is, but he has a new phone and the sound level hasn't been adjusted, so Siri starts yelling at a shocking volume in the otherwise quiet bog, causing both of us to recoil from robot-speak about selectively melting permafrost.

"Oh my god, that's so loud," Nathan says, and turns it off.

The actual definition of *thermokarst* is a series of marshy pools where icy permafrost thaws. On the discontiguous permafrost that undergirds much of this area, the water rises to the surface as ice melts deep beneath the soil, creating the effect we see. As temperatures warm and more ice melts, pools become ponds become lakes.

"That's what they say is happening here," says Nathan. We keep walking and find a vole marooned on a root, having swum across a part of the bog. Nathan points, crouches down. "There's so much more water. I just am looking at all those trees that are totally submerged and totally broken and dead." Nathan thinks these trees might have been on land last he saw them. Maybe wet. But certainly not underwater. He's been away the last year at Gustavus Adolphus College in Minnesota, hasn't been here in a while, and the changes are striking. Looking at how uneven this ground is, he doesn't remember that either. Nearby, a red squirrel chatters. "It really is like a ghost town, just a bunch of dead birch. They're just waiting to fall over, basically."

Along the boardwalk, there are educational signs describing the thermokarst, explaining why the lake is swallowing the forest and the trees are falling over. We stop to read one, and Nathan notices the logo in the lower left corner—ConocoPhillips. He reads it out loud: "Many scientists agree that the global climate has warmed in recent times, especially in the northern Boreal Forest." The sign goes on to say that if the warming trend continues, the boreal forest will change dramatically. "That's a super politically correct way to say that people are screwing with the climate," Nathan says, then ad-libs what the sign doesn't say: that humans are causing global warming and that when the forest changes, with it will come mass extinction and large-scale global die-off and an unstable future for children.

In the last year, so much has changed for youth and climate. New, daunting deadlines, worsening ecological conditions, and a rising up of youth to urge climate action. Yet so much is the same. A nation in denial.

Nathan looks the part of that guy who might not care. Six foot four and burly at nineteen, sometimes bearded and presidentially square jawed, he's a red-state resident who likes to hunt and fish and play with guns. But he is unique in his advocacy for this place, an outcast as a *Juliana* plaintiff, while Alaska is home to so much climate denialism and the unusual politics that attend it.

This summer of 2019, a long-shot Republican governor who was elected to preside over a budget crisis had been busy culling $444 million from the state budget, a huge chunk of which was to satisfy an absurd campaign promise to triple dividends to state residents from oil revenues at a time when oil prices were falling precipitously. Taxless as Alaska is, it did not have the funds to operate. And literally overnight, a galling $136 million disappeared from the University of Alaska budget, spread over three campuses. The immediate effect wiped out the scholarships of would-be students who had already packed bags and declined other universities, and the university's Board of Regents declared financial exigency, the first step in dismissing tenured faculty.

That Governor Mike Dunleavy received a master's degree from the very university he proposed to slash—and in a field he proposed slashing: education—was a deep study in irony. But no one had the time. At the Geophysical Institute in Fairbanks, scientists were too busy bracing for cuts while early plans looked to consolidate the three universities into one. It spelled trouble for the climate effort and especially for the Geophysical Institute—one of the preeminent polar research facilities in the world.

Seeing the thermokarst, these bits of the Alaska interior, makes it easy to understand why so many people come here just to study the earth. Launch a satellite into space over Alaska's polar region and you can see a whole lot of it. And better yet, those same satellites pass over Fairbanks fourteen times a day, making it easy to downlink the data they collect. This explains why NASA keeps satellite dishes at the University of Alaska in Fairbanks, more specifically at the Geophysical Institute, and why scientists from all over the world think of this region as a kind of living laboratory. Alaska has fifty-four volcanos that erupt every few months, plus a hundred thousand glaciers, a whole lot of sea ice, and some very strange weather. Ninety percent of the state and half of the Arctic stands on permafrost. And on that shaky land, Alaska experiences, on average, forty thousand earthquakes a year, a collection that includes three of the biggest earthquakes ever recorded.

Thus, the Geophysical Institute is one of seventeen university-affiliated research centers in America. And the State of Alaska similarly houses its key scientists there, including the state seismologist and the state's climatologist at the Alaska Climate Research Center. This latter center plays an important part in climate science generally, archiving records, developing statistics, and writing weather summaries for Alaska's newspapers. It's where scientists also compile information on Alaska's climate for whoever's asking, including scientists from all over the world. They compile information on everything from the physics and chemistry of the atmosphere to climate variability, the weather in the upper Arctic atmosphere, and how the atmosphere interacts with land and water and pollution. Which makes it critically important to polar science, and modern climate science too.

Climate change begins and ends with ice. It is where the story of the changing world can be fully understood. For the last hundred years, Earth's ice has been melting, reflecting less solar heat off the planet, causing it to warm. This melting ice makes the seas rise, changes weather patterns and ocean health, and all of those things promote species extinction and disease, and will eventually swamp land with water.

But this summer, while statewide rancor over cuts—to the university, Medicaid, support for people with disabilities and others—spread like stomach flu at a summer camp, the governor proposed killing off these research facilities, offered to reduce the cuts to $50 million in exchange.

Of course, what the governor proposed is illegal. Only Alaska's Board of Regents can make decisions about how university funding will be spent. But it didn't stop the governor from suggesting it anyway, even though the regents didn't take the bait.

Where climate breakdown is concerned, the governor's membership in the cult of deniability is well known. He was quoted in the *Anchorage Daily News* as saying, "The issue of global warming, in many respects, it's still being debated as to how to deal with it, what exactly is causing it. . . . I know there's a lot of folks and scientists who believe that man is contributing to this. But the question is, What is Alaska's role in this? What is Alaska doing?"

Sound familiar? It's a version of the same cookie-cutter quote offered by Koch Industries' Sheryl Corrigan at a 2016 event, the same quote offered by former secretary of the interior Ryan Zinke in response to the Camp Fire in California.

Indeed, the Koch-funded Americans for Prosperity had just sent Dunleavy on a controversial roadshow to talk budget priorities under the absurdly named Fortifying Alaska's Future series. This after the Republican-controlled legislature balked at his draconian cuts.

"There is no doubt in my mind whatsoever that the governor and especially the Koch brothers would love to shut down one of the world's largest Arctic climate change research facilities," Nathan tells me. We are standing at a viewpoint overlooking the thermokarst. In one direction is the deadened bog. In another, the boreal forest as it once was. Trees standing lush and tall. The earth soft but not submerged. Nathan says he remembers when this park looked more like the latter. And he is angry about what the governor's cuts could mean for the Fairbanks economy and to climate science.

From his Gold Hill home on one ridge to the University of Alaska at Fairbanks on the next, to West Valley High School just down the hill from that, where Nathan graduated in 2018, this is his world. In the same part of town are the coffee shops where he loves to sit for hours in deep, roving conversations with friends, many of them similarly high achieving and brainy. There's the place where he goes to haul water for 2.2 cents a gallon in the pickup his family keeps for such things. And somewhat in the middle of this, behind the Geophysical Institute at the university, running in either direction toward the various pieces of Nathan's life, are the miles and miles of ski trails he once trained on as a high school skiier. In case I need to understand how serious these sports are to his high school, Nathan explains that Olympian Kendall Kramer, who is one of the top three junior skiers in the United States, is a student there.

Like *Juliana* plaintiff Tia Hatton, Nathan has spent much of his life on skis. Tia is also a competitive skier whose last high school season in Bend, Oregon, was so snowless that she and her classmates trained on dry land and had to travel to places where there was actual snow to compete. When Nathan similarly competed, he trained twice a day, six days a week on these trails. There are dozens of miles of them encircling the Geophysical Institute, mostly gliding through timber but across fields and lakes too. He could ski endlessly if he wanted. But not without snow.

This vantage from his life in Alaska is such that Nathan can see climate change daily. So can every other Alaskan. Sea ice forms later now, allowing the erosion of coastal shoreline to the south, a circumstance that's forcing the relocation of entire villages, some the homes of Nathan's friends. The summers are cooler and wetter. And the shifting permafrost is opening sinkholes in Nathan's neighborhood, not to mention that some of his favorite glaciers are no longer glaciers and that wildfires are worsening by the year.

It's impossible to make a logical argument denying climate change in Alaska anymore. Yet some of Nathan's best friends are deniers. If he lived in a liberal stronghold, say some city to the south along the West Coast, he might be inclined to break ties with these people. The social fabric of such a place would help him do it. But Nathan thinks it's as childish to throw people away as it is to deny facts. He knows these last deniers don't want facts—they want their lifestyle instead. If we offer them something other than regulation, he says—clean energy jobs, a new future—he thinks it won't be hard to convince them to come along. "Climate denial from most people is a front for 'let's not change our economy.' And it's about fear," Nathan says.

In his tiny town of Ester outside of Fairbanks, the facts of the changing world of Alaska's interior are clear enough. Summers and winters are getting wetter and milder, and the winter rains are heavy and bring more chinooks, too, lukewarm winds that warm everything until it rains. The rain can freeze on snow in Alaska's climate, turning everything into an icy hellscape.

There isn't otherwise much that stops for the weather in interior Alaska. Snow days off school and work are not a thing, skis and snowmobiles being perfectly good substitutes for passable roads. A Fairbanks winter brings pastimes like dogsledding, even at the solstice when the norm is a scant three and a half hours of bona fide daylight. But this ice—it shuts down school and lots of other things. One recent storm caused a power outage for up to two weeks, marooning Nathan's family in their home for a week. Fortunately, the weather was what Alaskans call "warm," or about 20 degrees, so Nathan's family kept heated by feeding a wood stove and entertained by reading out loud, listening to the radio, and trying to make meals.

"I probably read a book about climate change," Nathan says dryly, consumer of nonfiction that he is.

Now, people are growing barley in Alaska. They can't grow wheat yet, but if things keep going this direction, maybe they will.

This kind of thawing is less a problem if you don't live on permafrost. But around Fairbanks, such regular thaws open sinkholes that suck up infrastructure—a parking lot at the University of Alaska is known to swallow whole cars—and also render runways useless and cause houses and buildings to shift. And melting permafrost releases methane, a greenhouse gas twenty-eight times as potent as carbon dioxide, an unsettling fact that attends the sliding earth. So it's not just the Old-West vibe that gives Fairbanks an air of having happened fifty years ago, not just the retro architecture or that the *Daily Miner*, the local newspaper, has downtown real estate a decade after the newspaper industry began its precipitous free fall across the lower forty-eight. It's more the sloping construction, the crapshoot of engineering that wears the buildings years before their time and gives Fairbanks a sort of eternal air, the sense of time much longer than a life can be, than even the life of a building can be.

For buildings and civilization, time is running out. Clear as ever to watch the polar ice melt before your eyes. And everything here—the sheer vastness of the land, the arid wind, the rivers slicing through the valley, sometimes overflowing—reminds a person how little control they have over any of it. So much of the Alaska interior is proof that nature is the boss of you. That regardless of whether people rove it, this land is bigger, older, more powerful than any silly little dominion humans believe they exert. To live here is to consent to being swallowed by the planet and its rhythms. To bow to those, and to the animals that share them with you, while you chop your own firewood and haul your own water. To understand that whatever tiny noises humans make from the moment they first breathe until the end of them are just good-for-nothing peeps, white noise in a land that doesn't care. And if you ever forget any of this, you need only walk away from town, drift away from the urban landscape, and see how long you can last.

For many Alaskans, the land provides as much as the rest. As much as the oil fields, which employ more than two-thirds of working Alaskans and paid $749 million in wages in 2016. The salmon that run the rivers, the halibut off the south coast, the moose and the caribou sustain a hunter's way of life. When the sun shines high twenty hours a day and sunset leaves an everlasting twilight, people fish their limits and stock freezers as wide as garage doors. Hens range from their insulated coops to peck in the yard. Berries grow like gangbusters and vegetables, too, in light that never ends, the growing season so unlike any other that the Alaska State Fair boasts a cabbage-weighing contest for the biggest cabbage that can be grown in any year. (In 2019, the winning cabbage weighed 138 pounds.)

All this makes Nathan very aware of where he comes from. And from what kind of life. The day we meet in summer 2019 in Fairbanks, Nathan's family has just added forty-five salmon to a freezer already stuffed with moose and caribou. He's just been fishing for Arctic grayling on the Chatanika River too. And because he hasn't taken his gun safety course to get his adult hunting license since starting college in Minnesota, he's gone to the shooting range in the last few days just to blast off a few shots while it is all that he can legally do. Among other outings are "popping squirrels," which means shooting them and harvesting the legs. He tells me, "If you're a good shot, you can get all four."

He wants this, all of this, preserved. And he says on our walk through the thermokarst that even though he likes to hunt and fish and sometimes just shoot things, he doesn't necessarily think that any of that should interfere with his desire to thwart climate change, or his belief that it is real and threatening and that it is coming for his way of life before a lot of others. That he has friends who disagree, friends who don't believe in climate change at all, is a thing that he values, not a thing he resists. He knows very well just from living here what it takes to communicate with people who don't share his ideas. And he also knows, thoroughly and deeply, that he would not have joined the *Juliana* case if he did not believe in trying.

"I think of my friends in this capacity as representatives of that side of America that is just still wanting to cling on to this economy, and I think that's a valid perspective if it's the skills that you have. I enjoy being friends with them because it keeps me grounded in the whole picture and not just the fiery-brand, Bay Area, save-the-earth people." Then he apologizes for being derogatory, assures me that he holds nothing against Bay Area people. Then he goes on, says his father is a teacher, his mother a school nurse, both publicly funded jobs. "I have to remember that my parents' salaries are paid for by oil. I come from a state where 90 percent of the general unrestricted funds in the state budget come from oil. It's a one-track pony. It's like Venezuela." He knows what threats a new economy brings. When he talks about climate remediation, he talks about how people from interior Alaska—people from rural communities everywhere in America—will find economic anchor in a world that tries to halt the climate breakdown. Without that, he says, we will be leaving vast numbers of Americans behind.

Last summer, Nathan interned with Senator Lisa Murkowski and befriended another intern who now works as Dunleavy's secretary. "She's the one who makes sure he eats his cheese sticks on time and goes to his meetings . . . so I am by no means living in an echo chamber," he says. Her father is the deputy general of the administration, so Nathan met the man and tried to have a conversation about his concerns. "Not that he would care. I'm just a kid," Nathan says, but he tried all the same.

What he took away from this talk were details of the man's life, of his relationships to his wife, his children, his church, and a bit of politics. Nathan knows people expect that as a radical, he will begin this conversation like an enemy combatant, but his goal is friendship first, politics second. He thinks hyperpartisanism is toxic, and he found lots to like about the man. That he is a lawyer, which Nathan aspires to be. That he's a jack-of-all-trades, built his house.

"I do know that this view of the world is incredibly privileged. I don't have an identity that is vilified. In fact, I fit in with many of these groups. I have the ability to go into the quote-unquote lion's den without being vilified. And I kind of think it is a personal responsibility to be a peacemaker." It is not the usual firebrand activism. But it is his.

He says the man offered perspective on the administration, about how his job is simply giving advice, and how ideological politicians tend to have ideological ideas. As he does in a lot of conversations, Nathan listened, said what the man could hear.

I ask Nathan what his friend, the man's daughter, thinks of his advocacy for climate, and about climate change generally, despite her work for the governor.

"She thinks it's a joke," he says. Then he turns and starts down the stairs to the sodden boardwalk, the tree trunks leaning over the trail.

Fairbanks is the home of the Museum of the North. It's where they keep the steppe bison mummy found near here forty years ago, and the tusks of mastadons and woolly mammoths too. They're among the animals that used to range on Alaska's plains back when Alaska had plains. This was about twenty thousand years ago, when the earth was colder, the sea level lower, and the land not so spongy and plagued by the kind of shrubs that foul the diet of an herbivore. Caribou were around then too. And muskox. They used to wander in herds across the Bering Land Bridge—the bit of land that once connected northern Alaska with Siberia.

The theory is that caribou and muskox have bigger hooves than the animals that didn't survive when the earth warmed. North America stopped being a block of ice and sea levels rose around Alaska. Steppe bison and horses, mastodons and saiga antelope all went extinct in the region when the Bering Land Bridge sank into the sea, plain became tundra, and food got scarcer right about the time the animals had to walk farther for it. Some just couldn't do the walking, sank into the boggy land instead, their diets unable to adapt to the plants. These were the end times for the Pleistocene epoch. Humans roamed then, too, lived once with the mastodons and the mammoths. They survived for the simple reasons of having the right diet and the right kind of feet.

CHAPTER SIXTEEN

November 4, 2019: Purposeful Procession

Everyone is looking forward to what the court will rule. But for months, there is nothing.

Then in September, Greta Thunberg arrives in America via sailboat, en route to the United Nations Climate Change Conference, set to take place in Chile in December. While in the US, she testifies to Congress alongside plaintiff Vic Barrett, who speaks of his Garifuna people while Greta submits an IPCC report into the record and urges elected leaders to "listen to the scientists."

Avery is in an overflow room with a bunch of other people when this happens, watching on the TV monitors. Before the committee, Greta and Vic, Jamie Margolin, and a fourth young activist are seated at a wooden dais, microphones pointed at the mouths of each. At Greta's end of the table, at least a dozen photographers are clamoring for the closest close-up of her. After a committee chair makes a speech about the failure of his own generation to address the climate crisis, a few other congressmen clear their throats for a while before the youth are given five minutes each to speak. Tens of thousands of fires have just burned across the Amazon, so Amazon forest protectors are seated behind them. They wear beaded headbands and sashes and are listening to a translation via headphones.

Here's what the young people say: that the world is promising them a future that isn't there, that careers and trips to natural places are a mirage, that their generation lives with fear and despair, that they face psychological impacts for life and are being discriminated against by a government that panders to corporations making billions off the destruction of their fates. "How do I convey to you what it feels like to know that within my lifetime the destruction we have already seen from the climate crisis will only get worse?" Jamie asks. After they patiently take turns testifying, urging Congress to take action, they answer questions while the congressmen prod. They are asked about their fear. Asked to comment on world politics. And lectured by Representative Garret Graves, a Republican from Louisiana, about such things as the "charade" of the Green New Deal, China's projected emissions increases under the Paris Accord, and how Americans make cleaner gas than Russians. In the committee room, there is decorum and quiet, process and procedure, professionalism, if hollow at times. Save for the point where Jamie calls inaction "cowardly" and "shameful," there is a sense that process is under way, action is being considered, that the US government is still working, even as half the committee chairs are empty, many representatives having found something else to do.

In the overflow room, however, what is happening is something different. When congressmen break in with questions for the youth—queries like: How can we address the climate crisis without regulation?—Avery says, "This one guy would get on his phone—he was just watching with us—and he'd search something and go, 'This person has accepted blank amount of money from the fossil fuel industry.' He would call it out." This knowledge, it feels like power to her fourteen years. It also disabuses her of any hope that our nation's leaders are there to act on her behalf.

The *Juliana* youth have been learning these lessons on Capitol Hill for years now, and this particular trip is no different. They are keeping brutal itineraries to meet with national leaders on climate change, running from the Senate to the House through the tunnels under the Capitol, back and forth through security. Even as they do, it is clear that much of the effort is wasted. Avery is buoyed by the people who are happy to greet them, happy to hear their concerns. "You get to meet really cool people who are actually doing something and actually do care," she says. Those meetings feel like wins. Then there are the losses. "I've definitely spoken with people who are not at all there and just super, 'Uh-huh, uh-huh, oh that's cool. Uh-huh, look what I'm doing. Oh, I'm not going to let you talk. I'm going to talk about me! Okay, gotta go! Bye!' Oh god, it's disgusting," she says. Like the meeting where an elected official booked half an hour with the *Juliana* plaintiffs, kept them waiting for most of it, then walked in, shot a selfie with Levi, and left.

They press on. In the days that follow, Greta's celebrity is so out of control that she turns up at things just to lure the media, then sits on the sidelines so that others can speak. She tries to do this one afternoon when the *Juliana* plaintiffs hold a press conference in front of the Supreme Court with supporting members of Congress, but the media will not leave Greta alone. Levi and another young activist have to body block Greta to create some space, and Levi ends up having to call security about a cameraman who won't back off. When the guy lies about his aggression, starts arguing with Levi and the other activist in front of the adults, Levi gets mad enough that he has to be carried off by Vic, who, as he extracts him, reminds everyone that Levi has only just turned twelve.

It's all a bit of a detour from what the plaintiffs have to say. But they say it anyway. In front of the whitewashed facade of the court, flanked by marble statues and other plaintiffs, Xiuhtezcatl and Kelsey take turns at a microphone with a handful of people in leadership. Kelsey speaks to the foible of her experience in Washington, DC, lets loose with the anger of how some of it feels. This last week of meetings, there were days where it seemed like she was the only adult in the room.

"I feel extremely let down. By history. By the decisions of our elders past. I feel . . . old. I'm twenty-three years old. I've been doing this for more than half my lifetime," she says. Cobalt blouse, brown boots, the sky behind her is a brilliant blue, the clouds puffy and idyllic. She says she'd rather be celebrating the mobilization of the youth climate movement, be glad to just be here together, but she can't. "I'm the oldest of my cohort of plaintiffs and in many of the meetings with members of Congress this week, it is very apparent that I'm the oldest as well. How shameful that we are in such a fragile system. As a planet, as a nation, as a society."

Privately, this feeling of aging out, it is wearing. The older she gets, the more Kelsey feels out of place advocating for youth. In several of her meetings on Capitol Hill, she found herself explaining climate science to staffers who were roughly her own age, interns and upstarts who already have the information. She can see through their maneuvers. Has become more strategic. Feels uncomfortable in rooms full of rowdy kids. And lately she's begun to long for the classroom, for working closely with children again, for getting back to the spaces where she feels most at home, not out of place among high-level thinkers and scholarly types. They don't feel like her people. And the critiques and asides, they are exhausting. The ones that let her know she's always being watched (on Facebook, "I saw you in the cafeteria with a plastic fork," or this one by email: "You need to cover up. You were wearing a tank top, not painting a picture of youth innocence").

But what she tells the crowd is that the older she gets, the more she feels a responsibility to people who are younger. And the more she feels that responsibility, the more she has to keep fighting, feels like she is begging to be heard. "We are talking about our lives. We are asking those individuals who seem to be putting guns to the foreheads of all youth, to not only not pull the trigger but remove your weapon." She sounds frustrated as she says this. But she keeps her voice strong. It is time, she says, for the adults to see the toll their indifference is taking. The burden the young now carry. And to stop asking the young for the sacrifice of childhood at the altar of begging for change.

That sacrifice, Xiuhtezcatl says, it weighs. On his turn at the microphone, black blazer over a T-shirt, he says when he met these other plaintiffs at fifteen, he felt a "unique sense of relatability with one another's stories." Before then, he didn't always have a community, didn't feel a sense of power. He holds his hand up briefly—the height he was when his advocacy started—and talks about the isolation of that work for himself and others, especially the times that it went unacknowledged because of race. Now his sister is eleven. Paints. Loves her kittens. "I look at her and I think about myself when I was eleven years old and even just then my entire life was absorbed by this work. Even at eleven, I was traveling all over the place, speaking, rallying, mobilizing." He wants her to have another kind of life. Wants to keep her and other kids out of this struggle for body, for life, for childhood, to do a thing politicians have been too afraid to do.

Maybe it will happen now that the voices of these *Juliana* youth are no longer so unique, Xiuhtezcatl says. "We are small pieces of a massive puzzle of an entire generation standing up in our streets, in our courts, in honor of our ancestors and those on the frontlines for generations to leverage this moment in history as the time when we change everything.... Whether we make it or not, is entirely up to each and every one of us."

Our Children's Trust has an office betting pool. Even the most pessimistic in the bunch bet that the court will rule in October. Then October comes and goes. With it, the first anniversary of the almost-trial. Still, there is no word from the Ninth Circuit about whether the trial can go forward.

By November, when no decision comes, the mood around climate action recalibrates yet again. Even the parents of the *Juliana* youth start to talk about direct action, about what might be necessary to spur actual, lasting change, if not through litigation and the courts. These are discussions at first. But then they start to become real. Not undertaken by the parents, but by other people who no doubt have similar feelings, who see nonviolent direct action garnering results overseas. On Halloween, I'd met with two organizers from a resistance group in Portland, Oregon, who had given me a heads-up about a protest soon to unfold on the Columbia River. Over the next few days, there were text messages and conversations. Then, the night of November 4, another text: an address and a time to arrive, 3 a.m.

I don't know where I'm going, but I go. I've been told that a group of activists intends to protest a ship headed for the Port of Vancouver, and when I arrive at the address they have given me, it turns out to be a parking lot near a public dock in Vancouver, Washington. People are quietly readying themselves in the dark. And after I step out of the car to get some basics from an organizer, I see Kiran hop out of another car, looking, by their own description, like a Power Ranger: white dry suit with stoplight red shoulders and shins, a red knit cap to match. The dry suit is unzipped and the top of Kiran's body is coming out of it like a banana out of the peel: Kiran in wool sweater, the arms of the suit dangling at the waist. Apparently there are a few base layers underneath. Kiran is prepared to be cold. It is 37 degrees outside, no doubt cooler on the water. And this night looks to be long.

Until a couple of hours ago, when an organizer let it slip, I did not expect to see Kiran here. They live three hours away in Seattle. But this is what it has come to now. From early days, Kiran has been encouraged not to get arrested. All of the *Juliana* plaintiffs are supposed to avoid getting arrested—it is part of the agreement they made when they signed on to be plaintiffs in the case, the agreement to behave themselves. Now Kiran is about to climb the pilings under a dock on the Columbia River and chain up to it. In another hour, they will disappear into a foggy night on a Zodiac toting a dry bag and a hammock, ringed by climbing rope and chain. This unexpected encounter, I'm not sure what to think.

For now there is just the time killing. There are six climbers in this parking lot, plus a handful of people who will soon provide support to them and to members of the media like me. I don't know it yet, but a dozen or more kayakers are assembling somewhere upriver, and boats are motoring in from every direction. One of the organizers is nervous: we are a spectacle. Get in the cars, he says. So we do.

What's about to happen here has to do with a bulk carrier called *Patagonia* that has been slowly making its way toward the United States from India. An international effort has tracked this ship while it travels. Greenpeace lent a hand to Portland Rising Tide, the local group whose members have been waiting for the boat to arrive, in keeping watch over it. There's been intel from the ground, too, though I'm not especially clear who provided this information, where they were, or what it was. First Nation groups from Canada and a water-based direct action group from northern Washington called Mosquito Fleet are also in the mix, with some of their members among those in the parking lot and in the kayaks.

The ship's cargo is what's at issue.

The vessel is carrying some of the more than 3.7 million feet of pipe that's been manufactured for the Trans Mountain Pipeline Expansion Project in Canada, which will boost the capacity of oil transport from the Alberta tar sands to the coast of British Columbia. The $7.4 billion expansion is projected to triple the 300,000 barrels of oil currently transported from Edmonton and will carry heavier oils with higher potential to emit greenhouse gas, making it a possible trigger of climate tipping points. Thus it is hotly contested. Kinder Morgan used to own this project but abandoned it amid opposition from First Nation and environmental groups and lawsuits from provincial and municipal governments. It's since been acquired by the Canadian government, which continues to fund the expansion despite public opposition.

Kiran and I sit in my car while the rest of the group divides themselves among a few others. I have a lot of questions. I'm not sure why Kiran is doing this, breaking rank with the *Juliana* cohort and behaving in public in a way they've agreed not to. But the deeper we get into this conversation, the more it's clear that as much as Kiran believes that when you want to make change you try all the doors, lately there is a feeling that the front door, the courthouse door, is locked. Kiran feels they have been standing at that door for four and a half long years while the system just teases them.

"Part of why I'm here is to just give them a little reminder that they can play with us in the system, but we don't have to stay in the system to have our voices heard," Kiran says.

Kiran is aware of what this kind of comment means, what this action means. Not just that they are about to get arrested and that this second protest arrest will likely confine Kiran to a life of working at nonprofits—there have been three rejected job applications recently, due to a prior arrest for trespass—but also that by letting this sentiment fly in the news, Kiran is using their celebrity so that this protest can be seen. This worked, after all, when Kiran was arrested with fifty-one other people when they blocked a train at a Shell and Tesoro facility in 2016. The subsequent press thwarted a potential expansion that would have meant large increases in oil trains traveling through Washington.

Now, Kiran's presence could bring more attention to the arrival of this bulk carrier and to the plan to spirit these pipes from one of the quieter Pacific Northwest ports north via rail. The celebrity that attends being a *Juliana* plaintiff has become a strategic lever that can be pulled for the right opportunities. And without Kiran's willingness to pull it in this case, the media eye will wander, zip from Trump to wildfire to Trump to wildfire and race right past the climate fight, past most of the action to stop the climate breakdown. Earlier today eleven thousand scientists issued a declaration that the planet is in a climate emergency and the world's people face "untold suffering due to the climate crisis" unless there are major transformations to global society. Still, I had trouble selling an article about this protest and the fact that the companies involved in the Trans Mountain expansion had to shop around for a port willing to accept this acutely unpopular bit of cargo, a political hot potato, before Kiran offered to stand in front of the story.

By the end of the day, several local newspapers and news stations will have failed to report that this port is shut down. The news will come a day later, after five people are arrested, as if it is the police activity that makes all of this relevant, the message of the activists and the action they have taken, the level of risk assumed, just humdrum crimes committed against the establishment instead of the other way around. The Port of Vancouver USA Commission recently promised to support renewable energy and avoid new fossil fuel terminals, so it seems ironic that this is the port to accept the cargo that will further open the tar sand export market. But this political slippery slope isn't deemed newsworthy locally either. The news industry's attitude toward the whole affair is that what is happening in the United States is merely something about a pipe, the situation in Canada irrelevant, and even though this combination of events could create a climate tipping point that greatly impacts us all, few people are taking the time to put it together.

We talk about this media circumstance while we kill the next hour in the car. And about other things. About the generation gap. About the roots of folk punk. About whether the English language needs nonbinary words for niece, nephew, aunt, uncle. About church and music. The environmental movement and its history. And then it is time. Kiran gets a text and we step out of the car into the frigid night air. Probably it is 35 outside now. It is very foggy—not a good night for boating—but there is nothing to be done about that. The other climbers and their support team are scattered among the other vehicles. One is a van with curtains where the climbers will suit up. The group cloistered inside of it seems to be potentially large, but I can't tell how many are in there in the dark. Like always, some are a little nervous about my presence, but I try to talk to them about why they are there anyway. Some talk to me, some don't. Some ask me not to use their names. Another gives me the name of an animal to be revised later if it becomes okay to speak. What he means is, to be revised if he gets arrested and his presence becomes known. At that point, there is nothing to lose.

Lydia Stolt says it is okay to write about her. She is nineteen. She is an experienced climber and tied the harnesses that the team will be using. She shows me the knots, the clips and ropes. At first she doesn't want me to use her name or this information about her—she is worried she will lose the scholarship she won for college—but then she changes her mind. When I ask her why she is here, Lydia says, "I fear for my future. It's zero hour and I can't watch the earth die around me. I don't want to be thirty and telling my kids that I didn't do anything."

We move the cars to another parking lot, quickly assemble on the pavement above the public dock, and the tension amps up quickly. There are boats lined up along the dock below us: a handful of Zodiacs, a skiff, and a sport boat. A security vehicle passes in the parking lot once, twice. The support team moves quickly to the boats. Assembled on the dock then are organizers and press—me and a photographer, also a freelancer who has just turned up—plus a couple of activists who plan to livestream whatever happens next. Some of the boats have distinct jobs: to ferry the climbers, to act as liaison with the police, and to ferry the journalists. The climbers are still huddled in the van, still making last preparations and staying out of sight until it is time to launch.

We're not doing anything illegal—just a bunch of people on a public dock in the night—but there's a sense that might not matter, that what's happening here could be interrupted in any number of ways. The driver of the sport boat says the fog is horrible, that boating in it is terrifying. It doesn't matter. There is no time to wait. The boats are loaded up with gear and food, life jackets dispersed, everyone with lamps and all-weather gear. When it's clear the boats are ready, a text goes up to the van. Encrypted, of course. All of these communications have been encrypted, messages that disappear shortly after they arrive.

In a few minutes, the climbers appear. I look up just in time to see them, equidistant from one another, swiftly moving across an overhead gangplank, a backdrop of streetlights. They walk with a uniform stride, each their own silhouette: figure in dry suit, helmet, headlamp, ropes and chains dangling. Each has a dry bag off the hip, a hammock strapped to their back. I might never forget this image, this purposeful procession, eerie in lowered heads, rounded shoulders. Resignation. I don't have a camera ready so I stare until it is gone. It is clear then how much what is about to happen is not the pastime of people eager to be agitators. Not clownish, no impulse in it. For all the contrary portrayal that might follow, that often follows such things, it's a moment that's been carefully considered by the people inside of it. It is the culmination of a lot of planning and a lot of work. There are backup plans, bailout plans, safety plans with the ropes and helmets. Every climber is carrying food and water, and yes, wearing diapers. Every situation has been carefully mulled. No one wants anything to go wrong here. But it could. There are dangers besides arrest. And in these last minutes, as Kiran moves swiftly past me, it's hard to suppress an impulse to stop them.

The climbers drop into a boat that whisks them quickly into the night. I load into the media boat. It is so foggy that there is nothing to see around us but the black river and the nearby light of a highway bridge, slight reflections on the water a couple of dozen yards around us. I can't see the opposite shore. Can't see the channel. Can't make out the pedestrian bridge that juts out over the water just a few dozen yards upriver. It occurs to me to hope that the pilot knows what he's doing. He says again how terrifying the route in was, but that it's better now. It is not very reassuring. I am carrying a headlamp and two flashlights so I offer to throw them on the water, but he tells me it will only illuminate the fog, make things worse, so we untie the boat and drive slowly away, fog light off the stern.

It's just after 6 a.m. and while the cold has been bitter since we arrived three hours ago, it is even colder on the water. The temperature of the air drops by a few degrees as we head toward the port terminals, the temperature of the wind falling by a few degrees more. The dock where the *Patagonia* will make port is not far. We are there in minutes. The ship is set to arrive in about an hour, so by the time we reach the terminal, six climbers are already making their way up the ladders under the pier—two to a piling, spaced along the dock—in spite of the lack of light.

Kiran moves slowly, each step absorbing attention.

"After every moment, I got something clipped and then I took a breath. Rest. Then I got something else clipped. And then I took a breath," Kiran says. "I'm not a super-confident climber." Ending up in the water is not really part of the plan. Kiran is an appreciator of action movie classics, was excited when Harrison Ford talked of attending the trial. Now *Mission Impossible* comes to mind.

Then the last step. The lock-in. The assembling of the hammock and the climbing inside. Kiran has been up since the morning of the day before, running on pure adrenaline and the vegan calzones that were supplied to the team. The middle-aged people in the effort really knew how to bring the food, Kiran noted, and they were glad for it. The wait had been lovely. A comfortable Victorian house with plenty to eat and beer. The climbers had smartly loaded calories the night before, as the *Patagonia* moved downriver and the deployment time kept shifting—3 a.m. to 4 a.m., back to 3 a.m.—while it wasn't clear when food might come again. Now the adrenaline is wearing off, the eyelids getting heavy, and to adjust to this new environment is absorbing too. The underside of the dock is loaded with barnacles. It is dark and creepy, and a person cannot help but wonder what else might be down there with the dank, sodden wood.

Lydia locks in below Kiran, so at home on the ropes that she doesn't climb quickly into her hammock, wants to be out in the elements, swinging, wearing a moose hat with antlers. She is immediately bored. Wishes she'd brought something to do. She starts talking. Kiran can't meet her there.

"I was just like, 'I am not on your wavelength right now,'" Kiran says later. "I just told her straight up, 'I might actually just need to sit here in quiet for a minute.'"

The sun is starting to rise. And underneath, kayakers are collecting in the water, filling the space between where the *Patagonia* will soon arrive and where the climbers hang in their hammocks, looking from the river like six tiny cocoons woven into the vast breadth of the dock, the dock hundreds of feet long, dozens of feet high.

Chatter begins overhead, workers arriving as the sun rises, and the heckling soon falls on Kiran's helmet—talk of urination off the dock, jokes about what the climbers will do if they have to take a shit, critiques of their gear, the bits of it and the kayaks that are made from plastic, as if no one knows these are fossil fuel products. None if it is conversation Kiran thinks they should engage. There is too far to go to connect. "If someone doesn't realize the gravity of the issue . . . and doesn't realize that we need to change the whole system . . . they're probably not going to understand anything else I have to say." Plus, the talk feels political. Like the polarization that is all over the news, gumming up the Internet with tribalism and vitriol, is hanging out on the dock too. As the workers continue to arrive in hard hats, and port security officers in yellow vests radio for police, for Coast Guard, the divide is a thing Kiran can feel. Down the dock, overhead of other climbers, a man stands and suddenly shouts, "Trump! Four more years!" He is loud enough to be heard over the boat motors and the expanse of the water below, a full-lunged effort for the whole of the river. But no one says anything in reply. Down below, it is as if life has briefly skipped a beat while people drag their attention from the situation at hand—tense and dangerous as it is—and from their singular focus on the ship soon to arrive, to yet another insertion of Donald Trump into daily life. It is as if the news cycle has grown legs, stood, and started walking around in the middle of things, yelling.

On the media boat, someone utters, "Well, they're related."

And that is all.

Kiran does not talk. Does not want to fight. Does not want this moment of resistance to be swallowed up by anger. Eyes fixed on the fog, now illuminated by a sun that will not pierce cloud today, Kiran has a singular need: to see the ship come into view. While the heckling continues overhead, an image comes to mind. It is an image frequently used by organizers, but in this context it takes on new significance. It is the image of small fish, many of them, forming a fish large enough to swallow the biggest fish of all. The arrival of the *Patagonia* will only reinforce this. When it finally clears the fog, kayakers all around and the climbers in their tiny hammocks, the sheer hulk of it, the behemoth of what they are fighting, is menacing. The *Patagonia* is nearly 600 feet long, topped with massive cargo hold hatches and derricks. As it moves into the channel at 8 a.m., the lateral stripe of its red-blue hull noses through the mist like a whale among minnows. A single letter of its name is half the size of a person.

"That was an image that will stay in my mind for a long time. It was ominous," Kiran says.

The kayaks seem very small. The climbers in the hammocks smaller still. Several kayakers paddle toward the *Patagonia* as it advances and remain there as the ship arrives broadside along the dock. Once it sidles up alongside the pilings, its enormous size alters the entire landscape. What once was an expanse populated by kayaks, Zodiacs, and the sport boat becomes just a narrow channel, the other boats squeezed into the in-between. By then the Coast Guard has arrived, the officers saying they just want to keep everybody safe. Soon there will be firefighters and police too. And the procession will move downriver to an alternate pier, one where a seventh climber will stealthily deploy and chain themselves in the narrow space between the top of a sheriff's boat and a crowd of dockworkers overhead, no room to go up or down, only a small fleet of kayaks between them and the *Patagonia*, foiled a second time. Law enforcement will issue commands then, declaring this even narrower channel a new jurisdiction, ordering evacuation, and the climber will be arrested with four others whose job it is, ironically, to talk to the police.

But before Kiran will climb down, knowing that the *Patagonia* has passed and there is nothing more to do, there will be the long standoff between the ship and the resistance. Long stretches of blaring foghorn. Of police orders. Of yelling dockworkers. And of mostly quiet resisters, paddling and motoring beneath the vulnerable humans chained to pilings in between. Kiran knows how much is being asked of the dockworkers, of the deckhands, to be thrust outside the normal order of things this way. To be kept from work, standing on dock and deck, and made to consider whether this pipe—this next installment of energy infrastructure—will tip the earth to uncontainable catastrophe. This kind of action, it takes everyone beyond their comfort zone. Takes people outside the ordinary plans of jobs and money and everyday exchanges that are based on the unquestioned assumption that things are all right in America. This kind of off-scripting, it forces people to know that perhaps things are not as all right as they seem.

But what is also knowable, Kiran says later, hanging there in the hammock, staring at the *Patagonia* edging ever closer, a tugboat pushing it toward the dock, toward the climbers, is that there is something stronger than the rote systems of our world. Stronger than the need for this ship to dock, for clocks to be punched, for workers to unload this pipe, and for trains to carry it northward so that this pipeline can be built and funnel the dirtiest oil on the planet around the globe.

"Their humanity is what I'm putting my bets in," Kiran says.

And the feeling that comes then—in knowing that other people, not the habitual machinery of progress, will decide the fate of their lives—isn't fear.

It is safety.

Epilogue

A panel of the Ninth Circuit Court of Appeals "reluctantly" dismissed the *Juliana* case on January 17, 2020, the judges ruling what legal observers suspected: that the court doesn't have the authority to tell the executive branch what to do about climate change, doing so would take decades of supervision, something that would violate the separation of powers.

Not all three judges agreed. In a lengthy dissenting opinion, California district judge Josephine L. Staton argued that courts do have the authority to protect the young in the face of climate breakdown, and should. In her remarks, she noted: "The injuries experienced by plaintiffs are the first small wave in an oncoming tsunami—now visible on the horizon of the not-so-distant future—that will destroy the United States as we currently know it. What sets this harm apart from all others is not just its magnitude, but its irreversibility."

Our Children's Trust appealed the ruling on March 3, 2020, asking that the full court of the Ninth Circuit review the case and send it back to the lower court for trial. That appeal is still pending.

Acknowledgments

This book would not have been possible without the *Juliana* twenty-one and their courageous journey. I cannot overstate what a transformative experience it was to work with them, nor can I aptly convey how outstanding, dogged, tough, unwavering, and yet consistently individual they all are. The level of self-sacrifice they have given their advocacy is a lesson in humility I only hope to have adequately chronicled. I am awestruck by the turns of luck and fate that led me to their story and deeply grateful for their trust in telling this piece of it.

I am similarly thankful to their parents, many of whom are appointment setters and gatekeepers alike. They endured my emails and my presence on hikes, at dinner, in their homes and camp—in other words, many places parents might easily decide a reporter does not belong. That they saw my involvement in their children's lives as an opportunity to empower them imbued me with a sense of duty that I felt an ambition to serve.

Along the way, I had the assistance of two people whose close relationships with the plaintiffs facilitated much of my reporting. They are Meg Ward, the former communications director of Our Children's Trust, and Dylan Plummer, the organization's former public engagement organizer. Meg especially has my deepest thanks for understanding the depth of the reporting I wished to bring to the *Juliana* case and for her belief in its value.

My thanks is also owed to the lead attorneys for the plaintiffs, Julia Olson, Andrea Rodgers, and Phil Gregory, who always had the patience to explain the details of *Juliana v. United States* to me. They spent considerable time doing so and shared much of their personal journeys with me along the way. Mary Christina Wood similarly offered her time and expertise in conveying the legal theories undergirding the *Juliana* case.

This book was vastly improved upon by Melissa Melo and Lily Lamadrid, both sensitivity readers who graciously read for ageism, making many astute observations about my treatment of youth voices and culture and providing powerful feedback on other issues. Charles Hudson, a Hidatsa tribal member, kindly reviewed my portrayal of Indigenous peoples and issues. He has, over many years, greatly informed my aptitude for reporting on both. Kalen Goodluck, a journalist with Diné, Mandan, Hidatsa, and Tsimshian roots, reviewed Chapter 12 for cultural competency and provided critical input into the style treatment of the cultural references herein. I am indebted to these wonderful humans.

My editor, Stacee Lawrence, deserves all the credit for making this book a book. It was Stacee who saw the value in telling the story of a thing that did not happen. That she understood the interest and import that the backstory of this case could have is singularly what made this project happen. Lorraine Anderson provided the calm, even-handed final edits that required as much logistical planning as red ink, and did so in times of calamity around the world. I was immensely pleased to have her company in the final days of this work.

The writing in between—from the proposal through each chapter—was attended and aided by the untiring members of Writers Anonymous, who never complain, even when I am fatiguing them. Despite the tongue-in-cheek moniker for this bunch, I cannot let their contributions remain anonymous: Jon Ross, Linda Wojtowick, Amanda Waldroupe, Shasta Kearns Moore, Jason Maurer, Kelsi Villarreal. I only hope I can repay their skill and kindness in something other than more drafts.

In the early days of this writing, David Wolman provided next-level thinking on story structure that helped me to see where this book was headed. Rebecca Clarren is excellent at making my reporting as easy to perceive as my cynicism. She provided structural feedback on a first draft and reordered the narrative for the emotional tolerance of the reader. AmyJo Sanders also read a first draft and provided enormous insight into the strength of the material, and coached me on how to bring those strengths into focus. Jon Ross and David Shafer provided an extra dose of line edits. My work is vastly improved by all of their attention to it. It is Bill Tarrant, West Coast editor at Reuters, to whom I owe thanks for this book's title.

I am so glad for the presence of all of these writers in my life and for their understanding and company in the unique challenges we face. It is the community they create that makes working in this industry worthwhile.

Thanks is also due to my agent, Jessica Papin, who makes my life tremendously easier and richer in all kinds of ways. Jessica showed enthusiasm for this book before I knew that I would write it. Her early reaction to the idea of a book about a canceled trial was largely what caused me to pursue it. She is also a darn good editor, and rolled her sleeves up on this project more than once.

My sincere gratitude also goes to Jonathan Logan for his support for my work through the Logan Nonfiction Residency. I owe him more than a martini and a black T-shirt, though I suspect he would settle for either. The time I spent at the Carey Institute for Global Good allowed me to read the court record of *Juliana v. United States* while people did things for me like make my bed and cook incomparably good food. Josh Freedman, Carol Ash, Carly Willsie, and many others made this environment the most conducive to good writing and good journalism that I have ever known, as did the companionship of my cohort. Every one of these individuals improved upon my ideas and my creative experience in undertaking this work. I am very, very lucky to know them and to have spent this time in their company.

The grant that allowed me to begin my pretrial reporting on the *Juliana* twenty-one came from the Society of Environmental Journalists and is the namesake of the late Lizzie Grossman, herself an intrepid environmental reporter whose forays into climate journalism took her all the way to the Arctic. This early support was essential and ultimately allowed me to stick with this story at precisely the juncture when many journalists turned away from it. I was humbled to do so in Lizzie's name.

Throughout this work, my number one writing companion has been my dog, Onyx, however disinterested. She is a great walking buddy and a better reminder of what it means to breathe.

My husband, Bjorn, supported all the rest—my time in the Logan residency, the early conception of this book, and the life changes required to write it. That this included such annoyances as letting quiet rank higher than emptying the dishwasher on the household list of greater goods leaves me indebted to his humor. He is also sounding board and editor, and a critic who will not forgive being bored—in another words, the best kind. He is a good walking buddy, too, and I am grateful that he walks through this world with me.

Appendix: About the Youth Plaintiffs

This appendix lists the youth plaintiffs in *Juliana v. United States* in the order in which they appear as litigants in their lawsuit against the federal government. These were their allegations and ages as of September 10, 2015, the date of their first amended complaint, the most complete public account of their claims. Details of how each plaintiff is specifically harmed by the impacts of climate change have been edited for brevity. Additional evidence and testimony have since been brought forth over the course of their litigation.

KELSEY CASCADIA ROSE JULIANA

Age: 19

Hometown: Eugene, Oregon

Litigating harms: To the freshwaters of Oregon for drinking, hygiene, and recreation. To food from marine animals. To food grown by small farmers in the Willamette Valley and in her family's garden. To snowpack for recreation. To participation in summer sports because of rising temperatures, algal blooms, and worsening wildfires. To her psychological well-being because she understands the impacts that will occur in her lifetime. To shared experiences with her future children.

XIUHTEZCATL TONATIUH MARTINEZ

Age: 15

Hometown: Boulder, Colorado

Litigating harms: To his spiritual and cultural practices, emotional and mental well-being because of impacts to forests. To his personal safety, property, and recreational interests through increased frequency and intensity of wildfires, drought, and declining snowpack. To the availability of water. To air and water quality and health as a result of fossil fuel exploitation in Colorado and planned export of fossil fuels from Colorado.

ALEX LOZNAK

Age: 18
Hometown: Kellogg, Oregon

Litigating harms: To his family farm from drought, rising temperatures, migration of forest species, and increased wildfire risk. To fish in local rivers. To water bodies and forests traversed by the proposed Pacific Connector Natural Gas Pipeline. To wildlife, including deer, elk, wild turkeys, and seafood, upon which his family depends for food. To his ability to raise children on his farm. To his allergies and asthma from worsening pollen. To snowpack and glaciers for recreation.

JACOB LEBEL

Age: 18
Hometown: Roseburg, Oregon

Litigating harms: To the use and enjoyment of his farm because of dwindling water supply, rising temperatures, and increased wildfire risk. To his career through harms to his farm. To his family's financial health from investments in irrigation systems to contend with drought and from impacts to conifer forests that they manage for financial benefit. To his spiritual practice and well-being through outdoor recreation. To snowpack and oceans for recreation. To seafood that he eats. To the aesthetic and spiritual enjoyment of his property and increased risk of explosions and wildfire from the proposed Pacific Connector Natural Gas Pipeline.

ZEALAND BELL

Age: 11
Hometown: Eugene, Oregon

Litigating harms: To recreation in the Oregon outdoors and on Oregon's river system because of reduced water levels and increased temperatures. To his ability to play sports because of increased temperatures. To his allergies from worsening pollen. To his family's financial health because of lost seasonal income from ski resorts. To recreation because of loss of snowpack. To drinking water in Eugene, which comes from snowpack. To coastal recreation and to seafood.

AVERY MCRAE

Age: 10

Hometown: Eugene, Oregon

Litigating harms: To recreation because of harms to nature and wildlife from drought, warmer temperatures, decreasing water levels, worsening forest fires, and algal blooms. To her allergies from worsening pollen. To vacation opportunities from beetle kill and associated wildlife impacts. To winter recreation from loss of snowpack, harms to beaches and oceans, and to seafood.

SAHARA VALENTINE

Age: 11

Hometown: Eugene, Oregon

Litigating harms: To water levels on the Mohawk River, where her grandparents live. To freshwater resources and beaches for recreation. To her health due to asthma because of worsening pollen and decreasing air quality because of wildfires. To snowpack for winter recreation and drinking water.

KIRAN ISAAC OOMMEN

Age: 18

Hometown: Eugene, Oregon

Litigating harms: To outdoor recreation because of decreased water levels and rising temperatures. To his diet through impacts to produce and seafood. To snowpack for outdoor recreation. To oceans and beaches for recreation. To his health because of worsening allergies and decreasing air quality because of wildfires. To the Florida Keys, where he visits family.

TIA MARIE HATTON

Age: 18

Hometown: Bend, Oregon

Litigating harms: To her ability to compete as a Nordic skier. To outdoor recreation as a result of drought, rising temperatures, and forest fires. To her psychological health because of the destruction of the wilderness around her home. To beach communities and mountains for recreation.

ISAAC VERGUN

Age: 13
Hometown: Beaverton, Oregon

Litigating harms: To the trees in his backyard, which died because of drought. To recreation and ecosystems because of drought, wildfire, and declining water flow. To snowpack for recreation, exercise, and emotional and spiritual benefits. To his health due to worsening asthma from increased pollen counts and from smoke from wildfires. To his ability to play sports because of rising temperatures.

MIKO VERGUN

Age: 14
Hometown: Beaverton, Oregon

Litigating harms: To the Marshall Islands, her place of birth, because of sea level rise. To recreation because of declining water levels and heat waves. To the fish, ocean life, and seafood on which she depends.

HAZEL VAN UMMERSEN

Age: 11
Hometown: Eugene, Oregon

Litigating harms: To recreation because of increased temperatures, low water levels, and abnormal seasonal variations. To the Oregon coast for recreation. To her future ability to enjoy activities that were once important aspects of her childhood. To seafood. To declining snowpack for recreation and for drinking water. To her health due to episodes of heat exhaustion.

SOPHIE KIVLEHAN

Age: 16
Hometown: Allentown, Pennsylvania

Litigating harms: To her education because of missed school due to extreme weather events. To her home because of hailstorms and floods. To participation in summer sports because of rising temperatures. To her future opportunities because of rising sea levels. To her ability to work, support herself, and begin a professional career while the nation's coastal economic hubs succumb to sea level rise. To her ability to grow food. To her health because of pollen allergies, extreme weather events, and intense heat. To the fabric of civilization and all living things.

JAIME BUTLER

Age: 14

Hometown: Flagstaff, Arizona

Litigating harms: To the Navajo Nation because of water scarcity. To her family's stability and financial well-being because of water scarcity and the cost of water. To her Navajo family because of the possibility of displacement. To her spiritual practice because of her diminishing ability to participate in land-based ceremonies. To the Kaibab National Forest because of pine beetle damage and escalating fire risk. To her ability to grow food. To her health because of worsening allergies. To her ability to preserve her cultural traditions and way of life into the future.

JOURNEY ZEPHIER

Age: 15

Hometown: Kapa'a, Kaua'i, Hawaii

Litigating harms: To his ability to farm, drum, fire dance, perform Halau Hula O Leilani, and recreate, walk along, and bike along the ocean because of threats to the Hawaiian Islands. To his food security because of ocean impacts and sea level rise. To his health, safety, cultural practice, and recreational interests because of sea level rise, flooding, storms, and other risks to the Hawaiian Islands. To his emotional well-being because of these risks. To the water supply, agricultural conditions, and thus his ability to remain in Kaua'i.

VIC BARRETT

Age: 16

Hometown: White Plains, New York

Litigating harms: To his emotional well-being because of the increase in superstorms in the Northeast. To his health because of rising summer temperatures, which limit outdoor recreation, and because of worsening pollen allergies. To his home because of rising sea levels and more frequent storm surges.

NATHANIEL BARING

Age: 15

Hometown: Ester, Alaska

Litigating harms: To his psychological well-being from witnessing climate change and understanding the increasingly severe impacts he will experience. To his ability to ski because of reduced snowfall. To his family and community because of power outages caused by ice storms. To his health because of allergies and asthma made worse by wildfires and a reduced ability to participate in summer sports. To the fish, animals, and ecosystems on which he relies for recreation and food. To lost recreation because of the retreating of glaciers.

AJI PIPER

Age: 15

Hometown: Seattle, Washington

Litigating harms: To his health from breathing exposure to the Carlton Complex fire while in the Cascades. To recreation, diet, and aesthetics from harms to fish, shellfish, and freshwater in Puget Sound. To his psychological well-being from seeing his home harmed by climate change. To forests from pine beetles and to New Mexico, where his mother is from, because of water shortages.

LEVI DRAHEIM

Age: 8

Hometown: Satellite Beach, Florida

Litigating harms: To his ability to grow food for lack of rainfall and increasing heat. To ocean beaches for recreation. To his well-being because of his exposure to rotting seaweed, a decline in sea turtles, increasing flesh-eating bacteria, and dead fish in his recreational areas. To his community's existence and real estate values because of sea level rise. To his health while worsening allergies lessen his time outdoors. To his psychological well-being while he fears for his future, his island's future, and the future of the wildlife around him.

JAYDEN FOYTLIN

Age: 12

Hometown: Rayne, Louisiana

Litigating harms: To her safety and the safety of her community, having experienced three hurricanes and more tropical storms in her lifetime. To her safety, her family's safety, and their property, bodily integrity, food security, recreation, and economic stability from rising sea levels, increased frequency of storm surges, flooding, and high winds. To her health because of impacts to the quality of her air and water by fossil fuel development. To marshes that serve as natural storm barriers from the dredging of canals for oil and gas exploration.

NICK VENNER

Age: 14
Hometown: Lakewood, Colorado

Litigating harms: To forests from pine beetle damage and increasing wildfires. To recreation because of damage to forests, low water flows that prevent fishing, loss of snowpack, and rising temperatures. To his fruit trees, garden, and food supply by hailstorms, drought, and pests made worse by climate change; and thus to his health.

Endnotes

All interviews and quotations in this book were recorded and transcribed for accuracy, unless otherwise noted in the text. Scene description comes from the author's own observations and from photographs of same, because the author has kind of a bad memory for such things.

It should be noted that the science used to describe climate impacts in this book is the broadest consensus science available on the date in which the recorded scene takes place. In other words, it is science that the US government either developed or accepted as fact between October 29, 2018, and November 4, 2019. Some of the science was significantly revised between the point at which a scene took place and the date of this book's publication. For example, in late 2019, the National Oceanic and Atmospheric Administration revised its estimates for projected sea level rise, predicting sea levels may rise up to 8.2 feet by 2100, compared with the EPA's previously estimated 4 feet, the level that Levi Draheim's community used in planning documents. This outcome would affect millions more people than the 13 million projected to be impacted by sea level rise in studies cited from 2016. And scientists at the National Oceanic and Atmospheric Administration did indeed tie climate change to increasing hurricane intensity in a study published in May 2020. Vanguard science by experts in the Juliana case predicted these findings. The author chose to illustrate the plaintiffs' impacts with the more conservative science as a means of underscoring the severity of climate impacts even when that severity is constrained by consensus. Only science available on the date of the recorded scenes is used.

Chapter 1

Observations of the rally at the Wayne L. Morse federal courthouse in Eugene on October 29, 2018, are the author's, and were fact-checked against the work of photographer Terray Sylvester, who was also there. Quotations from the speeches of Philip Gregory and Julia Olson, both attorneys for the plaintiffs, derive from the speeches themselves.

The term "no ordinary lawsuit" comes from the November 10, 2016, order from US District Court judge Ann Aiken, in which she denied the government's motions to dismiss the *Juliana* case.

Quotes from Kiran Oommen following the canceling of their deposition on October 20, 2018, were directly observed by the author, as were the coffee shop and the attending incongruent music. Kiran supplied background about their own activism both during this interview and in subsequent interviews until January 2020.

Kelsey Juliana's quotes are from interviews on September 25, 2018, at Amazon Park in Eugene and on October 29, 2018, at the courthouse rally. The timing of Greta Thunberg's protests at the Swedish parliament comes from the *Time* article "2019 Person of the Year—Greta Thunberg" by Charlotte Alter, Suyin Haynes, and Justin Worland.

Retweets of Our Children's Trust tweets by Ellen DeGeneres and Leonardo DiCaprio were observed by the author on September 13, 2016, and August 7, 2018. Information about media outlets' plans for extended trial coverage came from interviews with Meg Ward, former communications and youth engagement director of Our Children's Trust, between March 2018 and January 2020.

The description of Xiuhtezcatl Martinez's speech at age six comes from a YouTube video of the speech titled "Xiuhtezcatl—First Public Speech at 6 Yrs Old" posted by Earth Guardians. Similarly, Xiuhtezcatl's speech to the UN General Assembly is available via YouTube in the video "Xiuhtezcatl, Indigenous Climate Activist at the High-level Event on Climate Change" posted by the United Nations. Xiuhtezcatl's appearance on *Real Time with Bill Maher* can also be viewed on YouTube in the video "Xiuhtezcatl Martinez: We Rise | Real Time with Bill Maher (HBO)." Xiuhtezcatl's appearance on *The Daily Show with Trevor Noah* is available for streaming on Comedy Central via the article "Xiuhtezcatl Martinez—Taking on Climate Change with 'We Rise.'" General background about Xiuhtezcatl also derives from his book (with Justin Spizman), *We Rise: The Earth Guardians Guide to Building a Movement That Restores the Planet.* Xiuhtezcatl's Trump meme was observed on his Twitter account on August 12, 2018, and determined to be a fake in a phone interview with Xiuhtezcatl on November 9, 2019.

Information about the president's views on the Ninth Circuit Court of Appeals comes from the *Washington Examiner* article "Exclusive Interview: Trump

'Absolutely' Looking at Breaking Up 9th Circuit" by Sarah Westwood and from the analysis "In His Own Words: The President's Attacks on the Courts" published by the Brennan Center for Justice.

Legal perspectives on the Trump administration's use of the writ of mandamus come from Amy Howe's reporting for the SCOTUSblog, specifically "In a Letter, Government 'Suggests' Hold for Trial in Census Citizenship Dispute" and from the paper "The Supreme Court, 2018 Term—Comment: The Solicitor General and the Shadow Docket" by Stephen I. Vladeck. Vladeck's work examines the politicization of the solicitor general and is relied on for tallies of writs used during the Trump administration. Vladeck's essay "The Solicitor General and the Shadow Docket" in the *Harvard Law Review* provides additional context. Other cases affected by emergency writs are also from the above sources.

For details of the timing of Brett Kavanaugh's appointment to the US Supreme Court, the author relied on "Brett Kavanaugh's Nomination: A Timeline" by Sophie Tatum at CNN.

Quotes from Vic Barrett are from an interview at the Wayne L. Morse federal courthouse in Eugene on October 29, 2018.

The global fate of raccoons derives from an article by Karine Aigner for *National Geographic* titled "Raccoons Are Spreading Across Earth—and Climate Change Could Help." Information about species extinction was sourced from the *New York Times* article "Humans Are Speeding Extinction and Altering the Natural World at an 'Unprecedented' Pace" by Brad Plumer.

Kiran Oommen's lyrics are reprinted with their permission.

Chapter 2

Jacob Lebel's quotes and the description of his farm and farmhouse come from interviewing him there and touring the farm on August 29, 2018.

Narrative about the remedies the *Juliana* plaintiffs seek in terms of remediation from the agencies sued in their litigation derives from Jacob's explanation and was fact-checked against the court record, from which additional detail was culled. Additional information derives from an op-ed by Jacob in the *Globe and Mail* headlined: "Why I'm Taking My Government to Court on Climate Change."

Background on *Urgenda Foundation v. The State of the Netherlands* comes from the case summary by the same name on eLaw, the Environmental Law Alliance Worldwide website, and also from the *New York Times* article "In 'Strongest' Climate Ruling Yet, Dutch Court Orders Leaders to Take Action" by John Schwartz.

Details about rising global temperatures come from the article "Global Climate Change Vital Signs of the Planet" on the NASA website. Additional information about the impacts of climate breakdown come from the Climate Change page of the Global Issues section of the United Nations website.

Information about Arcosanti comes from the Arcosanti website and the YouTube video "The City of the Future Is Already Here," posted by *The Atlantic*.

The court record was also used to describe the plaintiffs' process of proving standing, specifically the arguments made by both the Obama and Trump administrations in their respective defenses against the *Juliana* litigants.

Information about the changing length of the Oregon fire season comes from a report by the Oregon Forest Resources Institute entitled "Impact of Oregon's 2017 Wildfire Season—Time for a Crucial Conversation." Additional detail about fire impacts in the West comes from the Associated Press article "Science Says: Hotter Weather Turbocharges US West Wildfires" by Seth Borenstein.

The text exchange between Jacob Lebel and Alex Loznak was provided by Alex via screenshots, with Jacob's permission. It has been edited for brevity, with redaction of personal information noted.

Description of the Pacific Connector route between Jacob's and Alex's farms derives from the author's travels, from project maps published by the Center for Biological Diversity in the press release "Lawsuit Filed to Stop 677-mile Ruby Pipeline and Protect Endangered Fish," from a map of the Pacific Connector published on the Jordan Cove LNG website, and from a map of the Pacific Crest Trail published online by the Pacific Crest Trail Association. Background about the Pacific Connector project and the proposed Jordan Cove liquefied natural gas terminal comes from the Jordan Cove LNG website and from the *Washington Post* article "Top Trump Adviser Calls for Reviving Controversial Natural Gas Project on Oregon's Coast" by Chris Mooney and Damian Paletta. Additional background about Gary Cohn, economic advisor to Trump, comes from his Wikipedia entry and was fact-checked against the original sources listed there. Additional background regarding the Pacific Connector and the Jordan Cove LNG terminal comes from the Federal Energy Regulatory Commission's December 17, 2009, order on the project.

Alex Loznak's quotes and the description of his farm, animals, and family home come from interviewing him there and touring the farm on August 28, 2018, and from subsequent interviews through January 2020.

Michael Gerrard's quotes derive from phone interviews on January 17, 2020, and February 5, 2020. Additional information about the Sabin Center for Climate Change Law and the Lamont-Doherty Earth Observatory was verified online.

The episode of *The Daily Show* featuring Parkland youth is available on YouTube as "Parkland Shooting Survivors School Congress on Gun Violence: The Daily Show."

Information about Brett Kavanaugh's record of decisions on greenhouse gas regulation comes from the Inside Climate News article "What Brett Kavanaugh on Supreme Court Could Mean for Climate Regulation" by Marianne Lavelle. The previously cited CNN article "Brett Kavanaugh's Nomination: A Timeline" by Sophie Tatum was also used to reconstitute the order of events in the chapter.

Chapter 3

Nick Venner's quotes and the description of his home and community come from interviewing Nick at his home on September 17, 2018, from subsequent interviews with Nick through March 2020, and from spending time in his hometown on September 16 and 17, 2018. Perspective from Marie Venner, Nick's mother, comes from speaking with her on September 17, 2018, in her home.

Material attributed to Nick's pleadings in the *Juliana* case was sourced from the court record.

Information about the 2018 fire season comes from the author's own observations and from the article "Colorado's 2018 Wildfire Season Is One of the Worst on Record, and It's Not Over Yet" by Kirk Mitchell for the *Denver Post.*

Description of pine beetle damage in Rocky Mountain National Park comes from the author's observations on September 18, 2018, and from a phone interview with Colorado State Forest Service entomologist Dan West on July 22, 2019. Additional background information about pine beetle impacts to Rocky Mountain National Park and other western forests comes from the US Forest Service website and the Colorado State Forest Service website, from the "2017 Report on the Health of Colorado's Forests—Meeting the Challenge of Dead and At-Risk Trees" from the State of Colorado, from the Union of Concerned Scientists' Climate Hot Map, and from the "Mountain Pine Beetle" fact sheet by D. A. Leatherman, I. Aguayo, and T. M. Mehall posted on the Colorado State University Extension website. Information about the average temperature at Longs Peak in Rocky Mountain National Park comes from weather averages compiled by Google. Additional details used in the description of tree pitch come from the Sciencing article "The Difference Between Tree Sap and Tree Resin" by Donald Miller.

Facts about straw usage and the straw boycott come from Brenna Houck's article "Why the World Is Hating on Plastic Straws Right Now," published by Eater. Details of the plastics found on beaches during worldwide cleanups come from the Ocean Conservancy report "Building a Clean Swell."

Information about the Drawdown Plan derives from the plan itself, available at Drawdown.org.

The twelve-year-old fellow Coloradan referenced in relation to her boycott of Home Depot is Haven Coleman, codirector of US Youth Climate Strike. Haven's rationale for the boycott was relayed to the author in a phone interview on March 10, 2019. Information about Home Depot's response to the CDP climate survey comes from Gizmodo's reporting on the same, specifically Brian Kahn's article "How Brands Think They Can Make a Buck Off Climate Change."

Chapter 4

Jayden Foytlin's quotes and the description of her home, the Atchafalaya Basin, and the L'eau Est La Vie camp come from interviewing Jayden and visiting the camp and the pipeline construction with her on October 9, 2018. Cherri Foytlin was present for much of the day, and Cherri's quotes come from conversation with her in the same context. Both Jayden and Cherri were reinterviewed with follow-up questions on October 29, 2019.

Information about the Atchafalaya Basin comes from the US Geological Survey's Natural Resource Inventory and Assessment System for the region. Details about the route, materials, and cargo for the pipelines described in this chapter come from the article "Bayou Bridge Testing Support for More Crude Takeaway to Gulf Coast" by Carolyn Davis for NGI's Shale Daily, from BakkenPipelineMap.com, and from maps available on the Water Is Life website.

Facts about the Louisiana floods of 2016 and climate change come from the *Guardian* article "Climate Change Made Louisiana's Catastrophic Floods Much More Likely" by Oliver Milman. Additional detail about Louisiana rainfall comes from the study "Synoptic and Quantitative Attributions of the Extreme Precipitation Leading to the August 2016 Louisiana Flood" by S.-Y. Simon Wang, Lin Zhao, and Robert R. Gillies.

Writing about the history of the Jennings oil field, and biographical information about Jules Clement, comes from the American Oil and Gas Historical Society's newsletter, specifically the article "Acadia Parish Oil Seeps Helped Discover the Jennings Oilfield in 1901," which is available online. Additional information about Spindletop derives from an article by the same name on the History Channel website. The government's assessment of the oil reserves in Louisiana was detailed in the US Geological Survey's Bulletin 429, titled "Oil and Gas in Louisiana with a Brief Summary of the Occurrence in Adjacent States."

Background on how oil is found comes from the Croft Production Systems website.

Figures on energy use equivalent to the 9 million barrels produced by the Jennings oil field at its peak come from the EcoSalon article "Gulf Oil Spill by the Numbers: 16 Different Ways to Understand the Disaster" by Susan Chaityn Lebovits.

Information about oil and gas facilities in Louisiana comes from the Louisiana Geological Survey document "Louisiana Petroleum Facts" from its February 2000 Public Information Series No. 2. Economic impact data for Louisiana comes from the Louisiana Mid-Continent Oil and Gas Association's website and the report "The Energy Sector: Still Giant Economic Engine for the Louisiana Economy—An Update" by Loren C. Scott for the Grow Louisiana Coalition.

Reporting on the oil and gas industry generally was greatly informed by Paul Roberts's book *The End of Oil*. The present-day description of oil and gas commodities and their relationship to the stock market comes from OilPrice.com, "Why Oil No Longer Rules the Stock Market" by Alex Kimani. Context of oil and gas investments as percentage of the S&P 500 comes from the Zacks article "Oil Companies That Trade on the Stock Market" by Tim Plaehn.

Information about the criminalization of pipeline protest comes from the Grist article "After Standing Rock, Protesting Pipelines Can Get You a Decade in Prison and $100K in Fines" by Naveena Sadasivam. Additional detail about moonlighting officers comes from the article "Louisiana Law Enforcement Officers Are Moonlighting for a Controversial Pipeline Company" by Karen Savage on The Appeal website. Alleen Brown's articles for the series "Oil and Water" on The Intercept were also critical to the author's understanding of law enforcement's relationship with pipeline construction operations and the criminalization of resisters, specifically "The Infiltrator" and "'The Scales Are Tipped': Emails Show Louisiana's Close Relationship with Oil Industry, Monitoring of Pipeline Opponents."

Detail about penalties for protest of critical infrastructure in America comes from the International Center for Not-For-Profit Law, specifically the US Protest Law Tracker on its website. Additional context was provided by Elly Page, as was her quote, in a phone interview on March 18, 2020. As of April 6, 2020, the district attorney for St. Martin Parish has not prosecuted Cherri Foytlin and other protesters for felony penalties available under new Louisiana state law, instead opting to issue fines.

Birds in the White Lake Wetlands conservation area were identified using the Louisiana Wildlife and Fisheries bird identification card for the area. Descriptions of the region are the author's own. Kinetica's barge terminal for liquefied natural gas was identified through the company's website.

Reporting on the portion of stimulus funds dedicated to alternative fuels comes from the report "Assessing 'Green Energy Economy' Stimulus Packages: Evidence from the US Programs Targeting Renewable Energy" by Luis Mundaca and Jessika Luth Richter. The author's description of the energy industry's reaction to the *Juliana* suit, including seeking standing and challenging the suit for two years, comes from the court record. ExxonMobil's spending on climate denialism is documented in the ZME Science article "9 Out of 10 Top Climate Change Deniers Linked with ExxonMobil" by Mihai Andrei and in the Greenpeace article "Exxon's Climate Denial History: A Timeline" on the Greenpeace website.

Information about Trump administration appointees David Bernhardt, Frank Fannon, Mike Catanzaro, and Sean Cunningham comes from the following articles: "DOE Coal Rule Might Help Client of Ex-Lobbyist Who Crafted It" by Benjamin Storrow and Hannah Northey for E&E News; "Trump Pick for State Department Energy Job Approved by Senate Panel" by staff for Reuters; and "Meet the Fossil Fuel All-Stars Trump Has Appointed to His Administration" by David Roberts for Vox.

The author's description of *Crawfish: The Musical* comes from a Facebook Live video of the performance.

Chapter 5

The explanation of the Ninth Circuit district courts' history in intervening in disputes through structured partnerships comes from an interview with Mary Christina Wood at the University of Oregon School of Law on June 13, 2009, and from her book *Nature's Trust: Environmental Law for a New Ecological Age*, as does background information on the Fish Wars.

Information about the history and geology of Oregon was gathered at the Oregon Historical Society Museum in Portland, with some details about environmental policies and current events from the author's own knowledge. The length of the state's coast was sourced from Google and Wikipedia. Facts regarding Nez Perce fishers come from the book *Undaunted Courage* by Stephen Ambrose. Information about the materials used by Native American tribes comes from work by David G. Lewis, specifically the article "Houses of the Oregon Tribes" on his website NDN History Research.

Details about Willamette Valley agriculture are reported from the US Department of Agriculture Statistics website, specifically the article "USDA's National Agriculture Statistics Service Oregon Field Office"; from the *Seattle Times* article "What's New, What's Novel in Oregon's Willamette Valley Wine Country" by Brian J. Cantwell; and from the Wikipedia entry on the valley, checked against original sources and reviewed for accuracy. Additional details come from the author's knowledge.

Zealand Bell's quotations derive from an interview at his home on September 26, 2018. Zealand's parents, Michael Bell and Kim Pash-Bell, were interviewed the same day.

Avery McRae's quotes and description of her home come from an interview with Avery there on September 24, 2018, as well as follow-up interviews through October 15, 2019. Avery's parents, Matt and Holly McRae, were interviewed with Avery on September 24, 2018, and Matt McRae was subsequently interviewed through October 15, 2019.

Details regarding mass extinction and the ocean impacts of climate change come from the United Nations report "Nature's Dangerous Decline 'Unprecedented'; Species Extinction Rates 'Accelerating,'" available on its website, and from the expert report of Ove Hoegh-Guldberg for the *Juliana* case.

Information about Hazel van Ummersen comes from Hazel and an interview with her at her home on July 25, 2019. Sahara Valentine was similarly interviewed on July 31, 2019, at Spencer Butte in Eugene.

Reportage about why the *Juliana* case was dismissed comes from the court record.

Observations of Halloween at the McRae home are the author's.

Chapter 6

The quote in the chapter's opening comes from Jeffrey Bossert Clark, given in response to the one question allowed while waiting for the elevator in the Mark O. Hatfield Courthouse in Portland on June 4, 2019, following the hearing before the Ninth Circuit Court of Appeals in the *Juliana* case there.

The author's analysis of the government's defense strategy comes from her own observation, from interviews with attorneys for the plaintiffs between October 2018 and January 2020, and from reading former Shell advisor Michael Liebreich's analysis of climate litigation in BloombergNEF, specifically the article "Liebreich: Climate Lawsuits—An Existential Risk to Fossil Fuel Firms?"

Ann Carlson's quotes derive from a phone interview on February 18, 2020.

Sources of figures on fossil fuel industry spending on climate disinformation are compiled in the endnotes for Chapter 12. Comments attributed to Chevron attorney Ted Boutrous derive from the article "Chevron's Lawyer, Speaking for Major Oil Companies, Says Climate Change Is Real and It's Your Fault" by Sarah Jeong and Rachel Becker, published on The Verge.

Background information regarding Jeffrey Bossert Clark was compiled through several sources, the first being the author's research on the government's PACER (Public Access to Court Electronic Records) system, looking closely at the federal cases for which Clark was the listed attorney. Also used were Clark's biography on the Department of Justice website and Clark's biography on and contributions to the Federalist Society website. Comments attributed to Clark's talk at the National Lawyers Convention convened by the Federalist Society in 2010 come from a video of the talk, available on the Federalist Society website, called "EPA: An Agency Gone Wild or Just Doing Its Job?—Event Audio/Video." Clark's commentary for PJ Media can be found on the organization's website in the article "Can the EPA Rely on UN Science?" Additional background comes from the Inside Climate News article "Senate Confirms BP Oil Spill Lawyer, Climate Policy Foe as Government's Top Environment Attorney" by Marianne Lavelle and also from the E&E Special Report "Trump's Top Environment Lawyer: Don't Chain Me to a Desk" by Ellen M. Gilmer.

The quotation from Mary Christina Wood is from an interview with Wood at the University of Oregon School of Law on June 13, 2009.

Information about public perception of climate change and fear of climate remediation comes from public polls and research. The referenced reports are "Climate Change in the American Mind" from the Yale Program on Climate Change Communication and the George Mason University Center for Climate Change Communication by Anthony Leiserowitz et al.; "Surveying American Attitudes toward Climate Change and Clean Energy" by Jon Krosnik for Resources for the Future; "The Ideology of Climate Change Denial in the United States" by Jean-Daniel Collomb for the *European Journal of American Studies*; and the poll "Is the Public Willing to Pay to Help Fix Climate Change?" by the Energy Policy Institute at the University of Chicago.

Subsequent reportage on pending expert testimony in the case comes from the expert witness reports and responses filed by experts prior to October 29, 2018. Quotes from the experts, save the passage from Mark Jacobson where the author indicates having posed a question to him, come from the expert reports in the court record. The fees charged by the experts were disclosed in their reports. The noted interview with Mark Jacobson occurred via telephone on December 5, 2019. The lawsuit involving Jacobson and other scientists was checked against court documents.

Greta Thunberg's full speech to the United Nations is available on YouTube in the video "Greta Thunberg's Full Speech to World Leaders at UN Climate Action Summit" posted by PBS NewsHour.

Background on the Scopes trial comes from the article "Scopes Monkey Trial Begins" on the History Channel website. Information about the media outlets planning extended trial coverage comes from interviews with Meg Ward, former communications and youth engagement director of Our Children's Trust, between March 2018 and January 2020.

The author's characterization of stakeholders in the *Juliana* case who filed friend briefs on behalf of the plaintiffs derives from the court record. Information about how the *Juliana* legal team would have presented the plaintiffs' case in court comes from a phone interview with attorney Julia Olson on October 26, 2018. Sophie Kivlehan's comments about James Hansen's public speaking style come from a phone interview with Sophie on September 5, 2018.

Chapter 7

Quotes and statements attributed to Levi and Leigh-Ann Draheim, as well as the description of the Satellite Beach ecosystem, come from an interview with both on October 10, 2018, in Satellite Beach, and from follow-up interviews with Levi on October 29, 2018, in Oregon and by phone on October 18, 2019. Wind speeds the date of our first meeting come from Weather Underground and the weather history for the date for Melbourne, Florida.

Information about how climate change influences storm frequency and intensity comes from the EPA document "What Climate Change Means for Florida" and the article "Global Warming and Hurricanes: An Overview of Current Research Results" on the website of the Geophysical Fluid Dynamics Laboratory at the National Oceanic and Atmospheric Administration. General information about hurricane season comes from the *Sun Sentinel* article "Hurricane Season Will Soon Enter Peak Period: Latest Forecast Released."

Information about sea level rise projected for Florida comes from the EPA document "What Climate Change Means for Florida," the expert report of Harold Wanless for the *Juliana* case, and the Sea Level Rise Viewer on the website of the National Oceanic and Atmospheric Administration. The City of Satellite Beach Sustainability Action Plan by Z. Eichholz and K. Lindeman was a reference for information about the projected effects of sea level rise for the community and the nature of the area's near-shore coquina rock reef. Additional detail about coquina rocks comes from the Florida State Park website. Other information about projected sea level rise for Satellite Beach comes from the slideshow "Assessing Municipal Vulnerability to Predicted Sea-Level Rise: City of Satellite Beach, Florida" by Randall W. Parkinson for the Space Coast Climate Change Initiative. General background about sea beans comes from the website seabean.com. Information about the city's application of sand to its beaches comes from the *Florida Today* article "Satellite Beach, Indian Harbour Beach Sand Project to Begin in December" by Jim Waymer.

Reportage on Hurricane Michael comes from the National Weather Service website, specifically the articles "Catastrophic Hurricane Michael Strikes Florida Panhandle October 10, 2018" and "Hurricane Michael PSH," and also from the FEMA website article "Hurricane Michael Recovery Resources—Florida."

Descriptions of the Satellite Beach sewer and stormwater systems come from Leigh-Ann and were fact-checked against the *Florida Today* article "'Be prepared:'

Hurricane Dorian to Test Water and Sewer Systems in Brevard" by Jim Waymer.

Additional information about manatees and their habitat in the Indian River Lagoon system comes from the Defenders of Wildlife website, the map "Critical Habitat for the Florida Manatee (*Trichechus manatus*) as Defined in the Code of Federal Regulations 50 Parts 1 to 199, Revised as of October 1, 2000," and the website of the Center for Biological Diversity. Mortality rates for sea turtles derive from Florida Fish and Wildlife Conservation Commission data on sea turtle stranding, available on the commission's website, and from Brevard County, in the article "Brevard Endangered and Threatened Species" on its website. Additional information about climate change effects on sea turtles comes from the report "Climate Change Impacts on Florida's Biodiversity and Ecology" by Beth Stys et al.

Details of Hurricane Michael's impact on Bay County, Florida, and on its children come from the report "Bay County, Florida, Long-Term Recovery Plan" by Robert Carroll et al. Additional context for the Florida Mental Health Act of 1971 (commonly known as the Baker Act), the law under which the state took 122 children into protective custody, comes from a user reference guide from the University of South Florida and the "Baker Act" article on the University of Florida Health website. Lise Van Susteren's projections of compound trauma from climate disasters come from her expert report for the *Juliana* case. Ken Chisholm's quotes come from a phone interview on October 7, 2019.

Levi's listed media interviews can be found online.

Chapter 8

Information about the window of time world leaders have to halt climate change, according to the Intergovernmental Panel on Climate Change, comes from the *New York Times* article "Major Climate Report Describes a Strong Risk of Crisis as Early as 2040" by Coral Davenport and from the report itself, which is called "Special Report: Global Warming of 1.5° C" and is available on the IPCC website.

Julia Olson's quotes and statements derive from interviews at and around her office on June 14, 2019, and October 16, 2019. Additional context was culled from her keynote address at the Public Interest Environmental Law Conference at the University of Oregon in 2018, a video of which is available online on the Our Children's Trust Facebook page. General information about NORAD comes from the NORAD website.

Statements attributed to Kelsey Juliana come from interviews and conversations with Kelsey between September 25, 2018, at Amazon Park in Eugene, and March 11, 2020. Scene detail from her speeches derives from Kelsey.

The description of Mary Christina Wood's theories and of atmospheric trust litigation, generally, comes from an interview with Mary Christina Wood at the

University of Oregon School of Law on June 13, 2009; from her book, *Nature's Trust: Environmental Law for a New Ecological Age*; from the article "Atmospheric Trust Litigation—Paving the Way for a Fossil-Fuel Free World" by Ipshita Mukherjee posted on Stanford Law School's SLS Blogs; and from the chapter on atmospheric trust litigation authored by Wood in the book *Climate Change: A Reader*, edited by W. H. Rodgers, Jr., and M. Robinson-Dorn.

Reportage on the first strategic meeting of attorneys filing the 2011 atmospheric trust lawsuits comes from the October 16, 2019, interview with Olson, as well as from a phone interview with Victoria Loorz on January 14, 2020; a phone interview with Alec Loorz on January 21, 2020; and a phone interview with Philip Gregory on December 5, 2019. Alec Loorz's quotes derive from the January 21 interview, as does certain background information about him. Other background information about Alec was provided by his mother, Victoria, in the January 14 interview. Information about the content of the meeting and subsequent case strategy comes from the Olson and Gregory interviews, as does additional detail on the initial meeting.

Alex Loznak's quotes come from the August 28, 2018, interview on his farm. Alex provided scans of the 1961 exchange between President Kennedy and Representative Clinton Anderson, as well as a scan of the printed article, which was written by John von Neumann. The article, called "Can We Survive Technology?" is also available on the *Fortune* website. Additional information about von Neumann was sourced from Wikipedia and checked against the original sourcing for accuracy. The documents that were identified as those demonstrating government knowledge of climate change on the butcher paper mural in Olson's office were photographed and independently verified.

Much of the author's writing about intergenerational justice comes from reading the following: Douglas Coupland's *Generation X: Tales for an Accelerated Culture*; Jeff Gordinier's *X Saves the World: How Generation X Got the Shaft But Can Still Keep Everything from Sucking*; the UPI article "Many Gen Xers Desolate as They Reach Middle Age, Study Says"; the study "In Times of Extreme Weather Events: Gen Z and Information Seeking about Climate Change on Digital Media" by Ujunwa Melvis Okeke; the white paper "Irony Politics and Gen Z" by Joshua Citarella; and the author's own mental archive of Gen X artifacts. Many movies were harmed in the making of this attitude.

The statistic detailing the percentage of Americans concerned about climate change by age comes from the Gallup poll "Global Warming Age Gap: Younger Americans Most Worried" by R. J. Reinhart. Historian and journalist Bill Kovarik was interviewed by phone on October 10, 2019. The white paper about Gen Z's flare for ironic humor is Joshua Citarella's "Irony Politics and Gen Z," available on Citarella's website.

Andrea Rodgers was interviewed by phone on October 4, 2019, and provided background information about herself and her family's legal history. Olson provided additional background information about Rodgers's hire.

Chapter 9

Plaintiffs featured in this chapter were interviewed on the following dates and in the following ways: Nick Venner by phone on May 21, 2019; Jayden Foytlin by phone on October 29, 2019; Alex Loznak in person in New York on March 22, 2019; Avery McRae at her home on October 15, 2019; Kiran Oommen at the Twilight Café in Portland on December 17, 2018; Aji Piper at Director Park in Portland on June 4, 2019; Levi Draheim by phone on October 18, 2019; and Isaac Vergun at the Tualatin Hills Nature Park on January 12, 2019.

Current events detailed in this chapter derive from the author's own database of news reports for November 2018. The data was compiled from the Society of Environmental Journalists' news digest of environmental stories, *EJ Today*, and includes 142 articles from a variety of sources, including *The Guardian*, the *Washington Post*, the *New York Times*, HuffPost, CBS, Reuters, The Hill, Inside Climate News, EnergyWire, Bloomberg, *The Nation*, E&E News, the Associated Press, *Newsweek*, *Reveal*, NPR, Greenwire, Vox, BuzzFeed, and local sources. Additional information about butterflies comes from the WWF website.

It was The Walrus that predicted "America's Next Civil War" in the article by the same name by Stephen Marche.

Information about the Trump administration's rollbacks to environmental policy prior to November 2018 comes from the *National Geographic* article "A Running List of How President Trump Is Changing Environmental Policy" by Michael Greshko et al. Further analysis of the relationship between climate trends and migration to the United States from nations to the south comes from the *New Yorker* article "How Climate Change Is Fuelling the U.S. Border Crisis" by Jonathan Blitzer.

It was Vox that reported on the Republican party's attention to climate change in the 2018 elections in the article "Governor Races Really Matter for Climate Change. Here Are the Ones to Watch" by Umair Irfan. Additional detail about the Green New Deal comes from the HuffPost article "Democrats' Green New Deal Wing Takes Shape Amid Wave of Progressive Climate Hawk Wins" by Alexander Kaufman.

Reportage on case developments in November 2018 derives from the court record and from a press-issued statement from Phil Gregory. Observations from the City Club of Eugene event are the author's. The quotation from Stephen I.

Vladeck is from his essay "The Solicitor General and the Shadow Docket" in the *Harvard Law Review.*

The elongated object headed for Earth came to attention via David Freeman's NBC article "Scientists Say Mysterious 'Oumuamua Object Could Be an Alien Spacecraft."

Chapter 10

Trump's "enemy of the American people" comments are documented in the *New York Times* article "Trump Calls the News Media the 'Enemy of the American People'" by Michael M. Grynbaum. Additional information about attacks on journalists in America and abroad comes from the Reporters Without Borders website, specifically its US Press Freedom Tracker. Additional context comes from the article "Trump's Attacks on the Press Empower Would-Be Dictators" by Pat Perriello for the *National Catholic Reporter* and from Suzanne Nossel's article "Trump's Attacks on the Press Are Illegal. We're Suing" for Politico.

It was a Gallup poll that established public confidence in media was at an all-time low on October 12, 2018. The poll is available through the Gallup website in the article "US Media Trust Continues to Recover from 2016 Low" by Jeffrey M. Jones.

Xiuhtezcatl Martinez was interviewed by phone on November 9, 2019. Additional information about his life comes from his book (with Justin Spizman), *We Rise: The Earth Guardians Guide to Building a Movement That Restores the Planet,* from the video "Kid Warrior—The Xiuhtezcatl Martinez Story" posted on YouTube by Yahya Yayato, and from the *Denver Post* article "'I'm in Love with a World That's Falling Apart': Meet the 16-Year-Old from Boulder Trying to Save the Planet" by Bethany Ao.

Ross Gelbspan's book *Boiling Point* provided context on climate disinformation and the role of George W. Bush in fostering climate denial. Quotes and statements attributed to Bill Kovarik come from a phone interview on October 10, 2019. Similarly, quotes and statements attributed to Jim Detjen come from a phone interview on October 17, 2019. Quotes and statements attributed to Joe Davis come from a phone interview on October 4, 2019. Quotes and statements attributed to Phil Shabecoff are from a phone interview on October 21, 2019.

Additional context on media coverage of the environment comes from background conversation with colleagues. Additional background about the history of the Society of Environmental Journalists comes from the SEJ website and the article "SEJ's Creation." Analysis of coverage of youth and climate comes from the *Columbia Journalism Review* article "How Teen Climate Activists Get—and Make—Climate News" by Abby Rabinowitz.

Information about spending on climate disinformation in the last twenty years comes from Robert Brulle's work, and writing on the same in the articles "Report: Fossil Fuel Industries—The Goliath of Climate-Related Lobbying Efforts, Spent Billions" and "Not Just the Koch Brothers: New Drexel Study Reveals Funders Behind the Climate Change Denial Effort," the latter by Alex McKechnie, both for the Drexel University website.

Contextual information about the panelizing and infotaining of news comes from the *Washington Post* article "How Breaking News Got Panelized: On Cable, Journalists and Pundits Increasingly Share Space" by Paul Farhi.

Reporting on the work of Naomi Oreskes comes from her book *Merchants of Doubt,* authored with Erik M. Conway; from the film by the same name; and from Oreskes's *Science* article "The Scientific Consensus on Climate Change." Additional detail about the signers of the Oregon Petition comes from the Associated Press article "Odd Names Added to Greenhouse Plea" by H. Josef Hebert. Information about the work of James Hoggan comes from his appearance on *The Green Interview,* available through the show's website. Information about DeSmog's Climate Disinformation Research Database, specifically the number of entries, comes from Ashley Braun, managing editor of DeSmog Blog. Details of specific entries are the author's observations.

Background on C. Everett Koop comes from his Wikipedia entry, with source material verified and links to his original papers reviewed. Tobacco industry tactics on disinformation come from the paper "Inventing Conflicts of Interest: A History of Tobacco Industry Tactics" by Allan Brandt.

Facts about other surgeon generals' positions on issues stem from the following articles: "Surgeon General Calls Climate Change a 'Serious, Immediate and Global Threat to Human Health' by Kate Sheppard for HuffPost; "Surgeon General's Warning: We Must Act on Climate" by Common Dreams for EcoWatch; and "Why Two Ex-Surgeons General Support the 'Juliana 21' Climate Lawsuit" by Richard Carmona and David Satcher for the *New York Times*.

The climate hashtags used in this section were top performers on Best-Hashtags.com.

Chapter 11

Information about Trump's early reaction to the Camp Fire and reaction from Gavin Newsom comes from several sources: the article "Trump Says Camp Fire Is 'Total Devastation,' but Hasn't Made Him Rethink Climate Change" by Alayna Shulman and Dianna M. Náñez for the Redding *Record Searchlight*; Jeremy B. White's article "After Trading Barbs All Year, Newsom and Trump Meet at California Fire Zone" for Politico; the *New York Times* article "Trump's Misleading

Claims About California's Fire 'Mismanagement'" by Kendra Pierre-Louis; and "Trump Says California Can Learn from Finland on Fires. Is He Right?" by Patrick Kingsley for the *New York Times.*

The Fourth National Climate Assessment is available online on the National Climate Assessment website.

Observations of Paradise, California, following the Camp Fire are the author's own, as are observations of the federal press conference in Paradise on November 26, 2018. All individuals quoted in this chapter were interviewed by the author in person in and around Paradise and Chico, California, between November 23 and November 30. Reporting on the number of damaged structures, people missing, and acreage of the burn is also original reporting by the author for Reuters. Additional context about the cause and forensics of the Camp Fire was provided by the *New York Times* article "'We Have Fire Everywhere'" by Jon Mooallem.

The quote from Sheryl Corrigan comes from the Grist article "Breaking: The Climate Is Changing Because a Koch Brother Said So" by Melissa Cronin.

As of April 6, 2020, twelve of fifteen ethics investigations of Ryan Zinke by the US Department of the Interior are still pending. Of the three that have been concluded, two complaints were substantiated: claims that Zinke allowed his wife to travel in government vehicles and that Zinke reassigned government officials without clear criteria.

Chapter 12

Quotes and statements attributed to Miko Vergun come from an interview with Miko at the Tualatin Hills Nature Park on January 12, 2019. Additional information about the Marshall Islands' climate crisis comes from the *New York Times* interactive project "The Marshall Islands Are Disappearing" by Coral Davenport and Josh Haner, and from the *Los Angeles Times* report "Marshall Islands, Low-Lying US Ally and Nuclear Testing Site, Declares a Climate Crisis" by Susanne Rust. The *National Geographic* article "Rising Seas Give Island Nation a Stark Choice: Relocate or Elevate" by Jon Letman provided additional crisis descriptions and solutions proposed by the Marshallese. Population information about the Marshall Islands is from the Marshall Islands article on the World Bank Data website. Information about the symposium on climate displacement for Hawaiians derives from an announcement of the same on the University of Hawaii website and from a Sea Grant Alaska web page for the event itself.

Reportage on Americans displaced by storms and by the Woolsey fire derives from the work of Carlos Martin at the Urban Institute, specifically his blog post "Who Are America's 'Climate Migrants,' and Where Will They Go?" on the Urban Institute's website. The number of people displaced by the Camp Fire comes from the *Washington Post* article "Forced from Paradise" by Frances Stead Sellers. Additional details about impending sea level rise come from the paper "Going Under: Long Wait Times for Post-Flood Buyouts Leave Homeowners Underwater" by the Natural Resources Defense Council and the report "Millions Projected to Be at Risk from Sea-Level Rise in the Continental United States" by Matt Hauer, Jason M. Evans, and Deepak Mishra, from which the quote comparing climate migration to the Great Migration is taken. Information about climate migration patterns comes from the Center for American Progress website, specifically the article "Climate Change Is Altering Migration Patterns Regionally and Globally" by Jayla Lundstrom.

Quotes and statements attributed to Jaime Butler come from a phone interview on December 5, 2019. Much of the description of "sheep camp" is Jaime's, but the author used additional resources, including a description of hogáns from the Smithsonian ebook *Navajo Houses* by Cosmos Mindeleff and detail about the Navajo-Churro breed of sheep, diet, and cultural significance from the Navajo Lifeway website, specifically the article "A Short History on Navajo-Churro Sheep." The description of shepherd life is also aided by the PRI article "Navajo Women Struggle to Preserve Traditions as Climate Change Intensifies" by Sonia Narang, and the accompanying short documentary. Additional information about Navajo weaving derives from the Britannica.com entry on the same.

Information about climate impacts on the Navajo Nation comes from the USGS report "Increasing Vulnerability of the Navajo People to Drought and Climate Change in the Southwestern United States: Accounts from Tribal Elders" by Margaret Redsteer, Klara Kelley, Harris Francis, and Debra Block. An accompanying video, "A Record of Change: Science and Elder Observations on the Navajo Nation," posted to YouTube by the USGS, was also used. For information about wildlife threatened by climate change on the Navajo Nation, the author referred to the Navajo Nation Department of Fish and Wildlife report "Climate-Change Vulnerability Assessment for Priority Wildlife Species" by the H. John Heinz III Center for Science, Economics and the Environment.

Reportage on the Navajo Nation's climate adaptation strategy relies on the document "Climate Adaptation Plan for the Navajo Nation" from the Navajo Nation Department of Fish and Wildlife by Gloria Tom, Carolynn Begay, and Raylene Yazzie; the report "Consideration for Climate Change and Variability Adaptation on the Navajo Nation" by Julie Nania and Karen Cozzetto; and from a phone interview with Janelle Josea, wildlife technician for the Navajo Nation Department of Fish and Wildlife, on January 9, 2020.

Information about the Navajo Nation's relationship to energy comes from the articles "Coal's Days in Navajo Country Are Numbered" by Benjamin Storrow for

E&E News, "Long-running Coal Plant on Navajo Nation Nears Its End" by Felicia Fonseca for the Associated Press, and the report "Navajo Nation Comprehensive Economic Development Strategy" prepared by Fourth World Design Group for the Navajo Nation Division of Economic Development. Data on the youth population of the Navajo Nation comes from Navajo Nation Community Profile on the University of Arizona website through its RII Native Peoples Technical Assistance Office.

Additional information about drought conditions comes from the *Navajo Times* article "Drought Takes Heavy Toll on Roaming Horses" by Larissa L. Jimmy, from the Associated Press article "Dozens of Wild Horses Found Dead amid Southwest Drought" by Felicia Fonseca, and from the Associated Press staff article "Over 100 Horses Apparent Victims of Drought on the Navajo Nation near Cameron, Ariz" on azcentral.com. Details about support for thirsty horses come from the *Arizona Daily Sun* article "In the Midst of Drought, Volunteers Bring Food and Water to Gray Mountain Wild Horses" by Emery Cowan and the Gray Mountain Horse Heroes Facebook page.

Background on the Garifuna people comes from the short documentary *On Our Land: Being Garífuna in Honduras,* produced by Remezcla. Details of coral bleaching come from the National Oceanic and Atmospheric Administration website in the article "What Is Coral Bleaching?" and from the expert report of Ove Hoegh-Guldberg in the *Juliana* case. Additional details of Garifuna culture come from the Global Sherpa website.

The host committees of Vic Barrett's testimony to congressional leaders were identified in a press release from the House Select Committee on the Climate Crisis, available on its website. Vic's testimony to the committee can be found on C-SPAN under the article headline "House Hearing on Climate Change." His op-ed in *The Guardian* is titled "Yes, I'm Striking Over the Climate Crisis. And Suing the US Government, Too."

Chapter 13

Observations of the Twilight Cafe and surrounding neighborhood are the author's. Lyrics for Geophagia's "No Hope" are available on the band's Bandcamp page on the *antifascist gardening collective* EP. Permission to reprint those lyrics was provided by Kiran Oommen, who wrote them. Statements attributed to Kiran Oommen in this section are from an unrecorded conversation in the Twilight Cafe on December 17, 2018, and from a subsequent interview in Vancouver, Washington, on November 4, 2019. Kiran's meeting with the ecumenical patriarch was verified by a photo on the Our Children's Trust Facebook page.

The comment attributed to Julia Olson was part of her speech at the rally at the Wayne L. Morse federal courthouse in Eugene on October 29, 2018.

Details of the Ninth Circuit court's decision to grant an interlocutory appeal in the *Juliana* case come from the court record.

Quotes and statements attributed to Aji Piper come from an interview with Aji in Lincoln Park in Seattle on October 21, 2018. Information about Aji's state case in Washington comes from the Our Children's Trust website and the supporting documents found there.

Reportage on the RAN Toronto 2009 Nixon campaign come from the RAN Toronto website and from the personal account of organizer Joshua Kahn Russell, available on his Wordpress blog by the same name.

Chapter 14

National news shows and articles featuring the *Juliana* twenty-one in early 2019 were the *60 Minutes* program "The Climate Change Lawsuit That Could Stop the US Government from Supporting Fossil Fuels," available on the CBS News website; the *Vogue* article "Do Americans Have a Constitutional Right to a Livable Planet? Meet the 21 Young People Who Say They Do" by Julia Felsenthal; and the *People* article "Kids Suing the US Government to Take Action Against Climate Change: It's a '911' Situation" by KC Baker.

Greta Thunberg's rise was observed in real time, but the author referred to the *Time* article "2019 Person of the Year—Greta Thunberg" by Charlotte Alter, Suyin Haynes, and Justin Worland to confirm the timeline of her activities. Thunberg's speech at the UN's climate conference in Katowice, Poland, is available on YouTube in the video "Greta Thunberg Full Speech at UN Climate Change COP24 Conference" posted by Connect4Climate.

Reportage on Greta's influence on European activists derives from the BuzzFeed article "A Huge Climate Change Movement Led by Teenage Girls Is Sweeping Europe. And It's Coming to the US Next" by J. Lester Feder, Zahra Hirji, and Pascale Müller. The author's characterization of Greta's influence worldwide comes from her own reporting, from the *New York Times* article "How a 7th-Grader's Strike Against Climate Change Exploded into a Movement" by Sarah Kaplan, and also from the *Time* article cited earlier.

Information about the formation of US Youth Climate Strike derives from the author's own reporting for Reuters, including in-person interviews with Haven Coleman and Isra Hirsi, cofounders of US Youth Climate Strike, on March 15, 2019, and a phone interview with Haven Coleman on March 10, 2019.

Reporting on the Zero Hour signature-gathering campaign for *Juliana v. United States* comes from a phone interview with Jamie Margolin on March 1,

2019, and from an interview with Meg Ward, former communications and youth engagement director of Our Children's Trust, on January 15, 2020. Additional detail about the content of the friend-of-the-court brief comes from the brief itself, available in the court record.

Observations from the Capitol lawn during the US Youth Climate Strike march are the author's.

Quotes and statements attributed to Alex Loznak come from an interview in New York on March 22, 2019. Alex's interview of Jay Inslee for MTV News is available on MTV's website under the article header "On Record with Presidential Hopeful Gov. Jay Inslee." The *New Yorker* article involving Alex is "Jay Inslee, Candidate and Eco-Dude" by Andrew Marantz. Coverage of Alex by the *Yale Politic* can be found in the article "People Before Partisanship: How Young Millennials Are Taking Charge on Climate Change" by Sherrie Wang.

Details of the film *Goldfinger* come from the film.

Aji Piper's testimony to the US House Select Committee on the Climate Crisis took place on April 4, 2019, and is available on the committee's website under the headline "Generation Climate: Young Leaders Urge Climate Action Now." Mentions of the congressional selfies and videos derive from real-time reporting of social media events by the author.

Details of the June 4, 2019, hearing before the Ninth Circuit court on the *Juliana* case derive from the author's reportage on the hearing for *The Guardian*. The hearing remains available on YouTube in the video "18-36082 Kelsey Rose Juliana v. USA" posted by the United States Court of Appeals for the Ninth Circuit. Observations from the Director Park plaza are the author's.

Chapter 15

Quotes and statements attributed to Nathan Baring derive from interviews with Nathan, beginning by phone on August 26, 2018, continuing in person in his Alaska community on October 11, 2019, and at the federal courthouse in Eugene on October 29, 2018, and again by phone on January 3, 2020.

The description of a thermokarst comes from the entry by the same name on the Encyclopedia Britannica website. Additional context for the Creamer's Field thermokarst was provided by a phone interview with Matthew Strum, a geophysicist at the Geophysical Institute at the University of Alaska, on August 11, 2019.

Background information about Governor Mike Dunleavy comes from the following articles: "How the Kochs' Americans for Prosperity Hijacked Alaska's Tax and Budget Hearings" by Don Wiener for the Center for Media and Democracy's PRWatch; "Mike Dunleavy and the Billionaire Koch Brothers Fiddle While Alaska Burns" by Bob Shavelson for the Waterkeeper Alliance; "Dunleavy Quietly

Nixes Alaska Climate Change Strategy" by Jes Stugelmayer for KTVA; "Alaska's Trumpian Governor Just Threatened the Health of the Entire State" by Sam Davenport for Vice; "Governor's Roadshow Visit Paid by AFP, Koch Brothers" by Diana Haecker for the *Nome Nugget*; "Dunleavy Budget Road Show Sponsored by Koch Brothers Group Requires Secrecy Pledge" by Dermot Cole for his self-named reporting project; and "Koch Criminal Justice Reform Trojan Horse: Special Report on Reentry and Following the Money" by Ralph Wilson for the Center for Media and Democracy's PRWatch. The quote from Sheryl Corrigan comes from the Grist article "Breaking: The Climate Is Changing Because a Koch Brother Said So" by Melissa Cronin.

Information about budget cuts to the University of Alaska and its Geothermal Institute comes from a phone interview with the institute's director, Robert McCoy, on August 16, 2019. McCoy also provided general information about the institute. Additional details about the institute come from its website.

Description of the Fairbanks and Ester communities derives from the author's reporting there. Kendall Kramer's state record and detail of her U18 ski season come from the article "The Amazing Alaskan Runner Who Doubles in Nordic Skiing" by Cory Mull for MileSplitUSA and from the US Ski and Snowboard website in the article headlined "Kramer, Schumacher Sweep 2019 US Junior Nationals." Facts about Tia Hatton come from an interview with Tia in Eugene on September 25, 2018.

Information about climate change and ice comes from the article "Why Are Glaciers and Sea Ice Melting?" on the WWF website. Facts about methane come from the *National Geographic* article "Methane Explained." General information about worsening climate conditions in Alaska comes from the following interviews with University of Alaska scientists: Martin Stuefer, atmospheric scientist, by phone on August 28, 2019; Regine Hock, glaciologist, in her office on August 9, 2019; Netti Labelle-Hamer, atmospheric and space sciences, by phone on September 3, 2019.

Information about the oil and gas industry's impacts on the Alaskan economy comes from the report "The Role of the Oil and Gas Industry in Alaska's Economy," prepared for the Alaska Oil and Gas Association by the McDowell Group.

Additional information about Alaskan culture and the history of its oil comes from the Explore Fairbanks Visitor Centers exhibits and from the author's original reporting. Facts about the Pleistocene epoch come from the Museum of the North exhibits at the University of Alaska at Fairbanks.

Chapter 16

Information about Greta Thunberg's travels to the United States comes from the Geographical article "Thunberg and Juliana Plaintiffs Strike Back Against US Climate Denial Machine" by Matt Maynard, as does information about testimony to the US House Select Committee on the Climate Crisis. Greta's testimony to the committee, as well as the testimonies of Jamie Margolin and Vic Barrett, can be found on C-SPAN under the article headline "House Hearing on Climate Change."

Avery McRae's description of her time on Capitol Hill, including her account of activities in the overflow space during the Select Committee hearing, comes from an interview with her at her home on October 15, 2019. Characterizations of Greta's celebrity come from Avery's observations in this interview and those of Levi Draheim in a phone interview on October 18, 2019. They also stem from the author's own observation of press activities around Greta's visit to the United States.

Comments made by Kelsey Juliana and Xiuhtezcatl Martinez on the steps of the US Supreme Court were broadcast via Facebook Live and are available on the Our Children's Trust Facebook page.

Information about the Trans Mountain Pipeline Expansion Project comes from the Trans Mountain Pipeline website; a phone interview with Heather Stebbings, spokesperson for the Port of Vancouver, on November 4, 2019; and the *National Post* article "Liberal Government Approves $9.3B Trans Mountain Expansion Project, but Critics Say It's Too Little Too Late" by Jesse Snyder. Details of the *Patagonia* bulk carrier are from the ship's listing on MarineTraffic.com. All other observations of the November 4, 2019, protest action at the Port of Vancouver, Washington, are the author's own. Quotes and statements attributed to Kiran Oommen come from an interview in Vancouver, Washington, on November 5, 2019, and a follow-up phone interview on November 6, 2019.

Lydia Stolt's quotes comes from an unrecorded interview in Vancouver, Washington, on November 5, 2019.

Epilogue

The Ninth Circuit Court's decision comes from the court record. Quotations from the judges come from the ruling.

Index